U0296025

王毅，先后在清华大学获得学士和硕士学位，在英国剑桥大学获博士学位。现任教于清华大学建筑学院，博士生导师，主持王毅工作室，为国家一级注册建筑师。曾在美国麻省理工学院（MIT）任研究员（2001），在意大利罗马大学做访问学者（1998）。
王毅曾多次获得重大建筑设计奖项，如中国建筑学会建国60周年建筑创作大奖（2009），教育部优秀勘察设计二等奖（2003）、三等奖（2013），建设部优秀勘察设计三等奖（2004）。曾发表中英文论文多篇，著有英文专著《一个世纪的变迁——20世纪北京城市结构》（*A Century of Change: Beijing's Urban Structure in the 20th Century*）。

Wang Yi got both Bachelor and Master degrees form Tsinghua University in China and Ph.D from the University of Cambridge in the UK. He currently, as a supervisor of doctorial students, teaches at Tsinghua University and, as a First Class Registered Architect of China, hosts Wang Yi's Design Studio as well. He was SPURS Fellow of Massachusetts Institute of Technology (2000) and Visiting Scholar in Roma University (1998).
Wang Yi has won several important architectural design awards, such as the Grand Prize (1st) of the Architectural Creation Award for the 60th Anniversary of China by Architectural Society of China (2009), the 2nd Prize (2003) and the 3rd Prize (2013) of the Excellent Design Award by Ministry of Education of China, and the 3rd Prize of the Excellent Design Award by Ministry of Housing and Urban-Rural Development of China (2004). Furthermore, He has published dozens of papers in Chinese or English and authored the English book of *A Century of Change: Beijing's Urban Structure in the 20th Century*.

本土出品 LOCAL CREATIONS

王毅建筑创作札记 SELECTED WORKS OF WANG YI

中国建筑工业出版社

目录 CONTENT

自序
Preface

建筑本来是天然具备本土属性的。本土属性首先表现为气候及地理的特征——不同气候和地理条件下的建筑千差万别；其次表现为人文及传统的特征——建筑是一定意识观念与文化背景的产物；再次表现为技术及材料的特征——材料和技术的不同造就建筑的差异。

在全球一体化、快速城镇化的今天，建筑的本土属性也在快速一体化，或者说在快速消失。建筑师受制于客观的经济、法规、技术等诸方面束缚，建筑设计行为不再是创造性的，而是变成生产性或经济性的。所以，现在的建筑大多缺乏带给人的感动。

当建筑师在地段现场毫无头绪，找不到任何灵感的时候，对建筑创作而言，这种感觉也许不是一件坏事。这时候我们需要做的是，低下头来，仔细感受土地的魅力。建造房屋不是顺便拿点东西往那里一放就好了，我们是在与土地进行“交易”。我们需要吸纳土地已经存在的东西，再用当代的思维和方法，奉献出我们创造的东西。

本土属性依然是今天建筑创作的源泉。本土特征需进行选择、提炼、抽象和转化，以符合当代要求。本土属性可以提供给建筑师某些设计思路和模式语言，但建筑师难能可贵之处是在此基础上作出某些提升和突破。我在每一次建筑创作中追求的就是多多少少的、那怕就那么一丁点儿的提升和突破。

2013 年 7 月于清华园

Architectures are naturally inspired by local attributes. These attribute primarily demonstrate in the climatic and geographic characteristics: architectures vary drastically by local climate and geography; secondarily in the traits of local humanism and tradition: architectures are partially composed of concepts and cultural backgrounds; and thirdly in the traits of local techniques and materials: technical and material diversion leads to different architectures.

In today's world, where globalization and urbanization rapidly take place, local attributes of architectures are also unifying or, perhaps, vanishing. Architects are subject to economical, legal and technical compliances. Architectural design is becoming more like a manufacturing and economical process instead of a creative art. Most architectures today therefore are uninspiring.

When architects feel clueless and without inspiration on project sites, the feelings may not be purely negative to the designs. What we need to do at the moment is to lower our heads and finely appreciate the charm of the land. Building is not about just piling materials up sites; it is rather a "transaction" with the land. We are to absorb all that from existing context around, and then contribute to our creations with contemporary thoughts and methods.

Local attribute is still a source of architectural creations today. Local conditions need to be selected, refined, abstracted, and converted to fulfill contemporary requirements. Local attributes offer architects design inspirations and languages, but the most estimable achievement is to break and advance the fundamentals of them. What I pursue in each of my architectural designs is to achieve any insignificant breakthrough and advancement possible.

July, 2013 at Tsinghua University

王毅建筑创作评论

Comment on Wang Yi's Architectural Creations

《台湾建筑》编辑部

Taiwan Architecture Editorial Office

王毅，在专业养成的历程中，有清华大学一脉相承的优良血统。在研究所阶段就以华人艺术宫(1991)赢得“中国首届建筑设计作品大赛”青年组首奖。在取得中国国家一级注册建筑师资格的同一年，又以庞大的纪念性尺度与对应轴线的简洁几何形式规划设计北京北中轴上的“中华世纪门”(1998)，获得设计竞赛的一等奖。除了上述两件荣获大奖的作品之外，1997年并时设计外交学院专家及留学生宿舍与北京电影学院逸夫影视艺术中心，后者陆续于1999、2002与2003年分别获得北京优秀建筑设计奖、WA中国建筑奖入围与国家教委优良设计评选二等奖。199~2003年期间，为世界自然遗产——四川黄龙与九寨沟，分别设计黄龙贵宾楼饭店（编者注：后更名为黄龙瑟尔嵯国际大酒店）与九寨沟国际大酒店。

在实务界，王毅是一位成功的建筑师，他的设计思维经由实际的建筑案例得到了实践；同时，在处理真实建筑设计工作中的省思与体会，反馈给他在清华大学建筑学院此等高等学府内的教学与研究工作。在教学的自我要求上，持续虚心学习且充满着传授知识并养成新一代的建筑专业人员的责任与使命感。在繁忙的设计与教学事务外，也从未忘记作为一个大学副教授必须经营的学术工作。1999年所发表关于印度建筑的研究报告。提供了相关研究极具价值的文献基础。

对于这位才华洋溢的建筑师与青年学者而言，他认为“灵感”是一个设计者重要的思维方式。我们可以借由探究其获得大奖的两项作品，试着追索这位建筑师在铺陈发展设计时的思维脉络。

首先，在面对“华人艺术宫”的基地条件上，面对都市的角地，选择了一个完整的正几何圆形平面形式对应街角，并将空间量需求较大的实验剧场配置于其中，同时解决都市街角美学形式与最大空间量体配置的双重问题。现代美术馆与二期的华人博物馆两性质不同的空间，则分别以扇形与矩阵的平面秩序配置。而艺术研究培训中心以私设道路切分，单独设立于基地主建筑群的东侧。在外显形式上，这个设计方案有着鲜明的西方现代主义风格，是一套符合逻辑的理性解决方案。然而究其背后的形式秩序思维，不难发现那对于传统中国哲学与形态的意识根源。文王八卦的太极分布以一种表层形式对应的操作，旋转压缩成为四分之一圆的扇形分割以对应基地的平面形状，并将卦象背后的隐喻涵意对应到建筑物的配置轴线上，例如正东轴线寓意龙之所在，艺术研究培训中心的轴线位于其上意味着中华文化兴旺发达。太极阴阳二元浑成隐喻着实验剧场中喜剧与悲剧的交叠演出，如此的叙事说理或许有些单纯，但在帮助设计的参考线定位与形式变化基准上，确实达成了形意相通的目标。

数年后的“中华世纪门设计案”，可更清楚地了解王毅空间形式的来源。诚如他对于建筑设计中的秩序，犹似对于中国传统文化中“礼”这个概念的理解　具有伦理与标准的观念。西方建筑的理性秩序在东方借助“礼”的观念，成为蕴含充沛哲学涵意的设计基础。中国传统天圆地方的宇宙观在这个设计方案中以具象的形式描述呈现，“方”的发展援用九宫矩阵排列的古代理想城市形式，作为轴线端点建筑的形式平面基础；“圆”的发展则取用古中国天文用的浑天仪形式的片段。以现代的桁架结构技术，架起尺度骛人的交错双圆拱。

对一位中国本地的专业建筑师而言，自幼对于中国传统哲学与形式的耳濡目染，累积了丰富的东方文化基础，成为意识与思维不可或缺的一部分；完整且优良的科班训练，造就熟捻运用现代西方几何建筑元素的能力。民族的自信心是造就设计概念与说理，以一种当仁不让的态度，在这被西方思维几乎彻底淹没的时代，以一个充满差异的玄妙角色，赋予“西学为用”这具象物质建筑空间形体，一个神秘且满富内涵的“中学为体”意义。而这属于今日东方巨龙的“灵感”，将造就当代中国建筑的现代化。

原文刊于《台湾建筑》2004年第四期，题为《凌空飞跃的东方巨龙》

Wang Yi, in the course of his professional development, has exhibited the fine influences characteristic of the Tsinghua University family. Even in the midst of his education and research, he found time to exhibit his prowess in a design competition for the Chinese Art Palace (1991), resulting in a first place award in the youth group category. In the same year which he attained his 1st Class Registered Architect of China, he was also awarded the grand prize for his design of the China Arch for New Millennium (1998), a monumental structure composed of simple geometric forms located on Beijing's northern central axis. In addition to the two award winning projects above, he is also responsible for the designs of the China Foreign Affairs University Experts and Foreign Students Dormitory in 1997 and the Beijing Film Academy's Yifu Film and TV Arts Centre in 1999. In 2002 and 2003 respectively, he has been awarded the Beijing Architectural Design Award for Excellence, was a finalist for the WA Chinese Architecture Award, and was awarded the Second Place State Board of Education's Outstanding Design Award. In the period between 1998 to 2003 he has also partaken in projects on United Nations World Heritage Sites in Sichuan, such as the Huanglong Seercuo International Hotel and the Jiuzhaigou International Hotel.

In the professional community, Wang Yi is a successful architect, and his designs exhibit serious contemplation of architecture in actual practice. Simultaneously, his work in the professional environment has also provided him with invaluable feedback for his role as an educator and researcher at the Tsinghua University School of Architecture and other serious institutions of higher education. He has high standards, and continues to impart knowledge upon his students with the goal of building a new generation of professionals with a sense of responsibility and purpose. Outside of his professional and educational affairs, it is important not to forget Wang Yi's role as a researcher and scholar. His report on Indian architecture (published in 1999) provides valuable literature for foundations researching related topics.

In considering talented architects and aspiring young scholars, he believes that "inspiration" is an important factor in the contemplative process for designers. From the analysis of his two award winning projects, we can perhaps document this architect's train of thought.

First off, when considering the elementary site conditions of the Chinese Art Palace, the project faces the city diagonally, so a solution which corresponds to the shape of roads by utilizing regular geometric forms would address the dual problems of the urban corner aesthetic space as well as the relatively large spatial demand of program. The Museum of Modern Art and the second phase of Chinese Art Palace intrinsically define two different properties of space, so they are differentiated in the plan sequence by sectors and matrices. The art research training centre is segmented by a private road, so as to set up a separate eastern side for the base of the main buildings. In appearance, this produces a distinctive western modernist style, a result of logical and rational solutions. However, the roots of order and form behind its conception are obviously drawn from tradition Chinese philosophy. The idea of Emperor Wen's Eight Diagrams of Divination are corresponded to the design of the surface, a rotating compression creates a ninety degree angled fan shape which correlates to the shape of the foundation in plan. The metaphorical axis also relates to the configuration of the building, for instance wherein the dragon symbolizes the eastern axis. Therefore the placement of the art research and training centre on this axis allows for Chinese culture to thrive. The sun's *yin* and *yang* metaphorically hints at the experimental theatre's overlapping performances of comedy

and tragedy. Perhaps the narrative account here is a bit simple or shallow, but in terms of design, formal elements have become successfully interlinked with deeper philosophical meanings.

A few years after the design for the China Arch for New Millennium, the source of Wang Yi's spatial conception is much more easily understood. True to his understanding of the sequencing inherent within architectural design, his understanding of ethics and standards relate to the idea of *li* (rite) in Chinese society. Taking the rational orders of western architecture and borrowing the idea of *li* from the east allows for abundant philosophical meanings in a basis of design. This project's formal aspect borrows directly from the Chinese cosmology of "round sky, square earth". The development of the square form invokes the idea of the *jiugong* (nine equal boxes), arranged in a matrix representative of the ideal ancient city, becoming the plan basis for the architectural terminal point of the axis. The development of the round form is related to Chinese astronomy and fragments of the armillary sphere which describes celestial organization. This is done with a modern truss structure, allowing for an amazing staggered double arch of sizable proportions.

As a Chinese professional architect, subtle influences at a young age taken from traditional Chinese philosophies and forms provide a rich cultural foundation, one which later becomes an integral part of design that conception cannot live without. This is coupled with excellent training which allows for the cohesion of familiar geometries with western architectural elements. A people's self-confidence trains design conception and argument, resulting in uncompromising attitudes. In this era of thinking which is almost totally submerged by western thought of unclear roles full of differences, "western thought for application" is given denoting a figurative material of architectural spatial form, a significance which is mysterious but full of connotation emphasizing "eastern thought for the body". This, in effect, is what can be considered as today's eastern soaring dragon: "inspiration", which will bring about the modernization of contemporary Chinese architecture.

Originally published in *Taiwan Architecture*, 2004 (04), with the title of A Soaring Eastern Dragon

王毅的方案
Wang Yi's Designs

曾昭奋
Ceng Zhaofen

王毅，因在一次全国设计竞赛（华人艺术宫设计）中夺冠而引人瞩目。后来，当他的这个获奖方案在不同情况下多次被仿效、被移植的时候，就常常有人记起、提起：啊，这是王毅的方案！

1991年，王毅在题为“华人艺术宫”的全国建筑设计作品大赛中荣获青年组第一名。1998年，王毅的中华世纪门方案获竞赛一等奖。同年，该方案又被选为“北京建筑设计十佳”之一，再次获奖。

作为一个著名高等学府的建筑设计教师，并不是一个轻松的差使。为了做好教学工作，王毅的肩膀上压有两副重担：

一头是，作为一个青年建筑师。他经常参与建筑设计工作，从接受任务、方案构思、施工图绘制到现场落实。他说，如果自己不会做设计，不会绘制施工图，没有现场经验，在辅导学生的课程设计时就会觉得心虚。四川九寨沟国际大酒店将于今年上半年落成开业。王毅毅然投入这个他称之为既具挑战性、又具学术价值的设计项目。1994年，他在云南丽江出了车祸以后，多年不坐山间或乡间的长途汽车。但为了九寨沟宾馆，他重又坐汽车进了山沟，进了少数民族地区，去寻找川、藏民居特色和创作的灵感。

另一头是，作为一个年轻学者。多年来，他主动承担了一些科学研究工作。其中有一项是印度建筑研究，并于1990年发表了第一篇研究报告“香积四海”（参见《世界建筑》1990年第六期）。他曾经作为罗马大学访问学者，遍访欧洲各国。有了对中国建筑和欧洲建筑的认识，再去认识印度，就会显得更为清醒、更为冷静了。1999年，王毅在访问印度之后所发表的第二篇印度建筑研究报告（参见《世界建筑》1999年第八期），对印度建筑作了更准确的评析和论断，在国人的同类文章中后来居上。

在阅读王毅两个获奖方案的时候，我的一个始终不能淡化的印象是：他是一位教师，一位建筑师，一位青年学者。我相信，在它们的构思过程中，必然会有更多的说法。

我曾说过，设计竞赛是青年建筑师们最盛大的节日。获得过两次大赛桂冠和其他多项奖励的王毅则有自己的切身体会：“建筑设计竞赛是建筑师表达自我、展示才华的运动场，是新思想、新观点产生的摇篮。”

“节日”、“运动场”、“摇篮”，有一个相近的意思，她孕育、激励设计者创新的勇气，必胜的信心，冲刺的毅力，启发设计者的灵感。

灵感是什么呢？王毅曾引用钱学森院士的话：“灵感也是一种思维方式，灵感思维是与形象思维、逻辑思维并列的，现在人们对它知之甚少，但却是最具创造力的思维，一旦人们掌握了灵感思维的机理，将会大大激发人们的创造力。”

上述两个获奖方案的诞生，正是建筑师逻辑思维、形象思维加上灵感思维的产物。建筑师的、老百姓的和传统哲学家的观念和期待，都成为罗辑思维的凭借和内容，如太极图、八卦、城市广场、中轴线、天坛、天体运动、钢构拱门以及丰满、文明、伟大等，都是可以分析和言说的。而灵感和灵感思维所促成的建筑意境和形象的创造和定格，则是属于建筑师自己的，似乎是只可意会而不可言传了。

文彦曾回忆20世纪50年代在哈佛大学听路易·康关于他设计耶鲁大学美术馆（1951~1953年）的演说。康说，当他发现了最重要的中心理念时，正在着手构思、设计中的方案的各部门如平面、立面、细部大样及其他相关部分都极其自然地一气呵成地大功告成。矶崎新认为，中心理念就是建筑师的构想(Concept)，有“受胎”、“孕育”的涵意。这种中心理念、构想、受胎和孕育是否可以理解为建筑师产生灵感和灵感思维的过程(灵感一旦出现，一通百通，一切迎而解)?我想让王毅以他的两个获奖方案为例，对灵感和灵感思维作一些现身说法。

精妙的构思不是简单与简单的相加，而是相撞！简单与简单的巧妙碰撞才能产生几何倍数的感染力，这里需要的是灵感和灵感思维。灵感思维所孕育的创造力是无比巨大的，而灵感的产生需要长期的知识积累过程。许多建筑设计竞赛对于建筑学的发展起过巨大的推动作用，其中反映的新思想和新观念是至关重要的。建筑的内部逻辑和外部条件提示了建筑的若干可能的基本形式，然而建筑整体形态的产生却是一种超越。而正是这种超越使建筑不仅仅是一种构筑物更是一种文化财富。另外，建筑创作从构思到完成是一个极其艰难的过程，其中的匠心有时就似一个“黑匣”，很难把它解开说得一清二楚。这其中就包含了许多个性化的因素，与建筑师的人生倾向、艺术修养有很大关联。

灵感是一种个性化的、创造性的思维活动，它能够跃进，达到逻辑思维和形象思维所不能达到的另一种境界，建筑作品的成败高下，就取决于此。安于模仿、抄袭别人作品的建筑师和艺术家，虽然有相似的知识、动机和愿望，但却缺少这种灵感和灵感思维活动。

作为一个教师，王毅可以自己丰富的知识、思想、经验等，指导他的学生进行逻辑思维和形象思维，至于灵感和灵感思维的突发与成功(十月怀胎、一朝分娩的过程)，则暂时只能从建筑师的精品包括王毅的两个获奖方案中，去体验、捕捉和猜测？

原文刊于《新建筑》2000年第一期

Wang Yi first came to prominence through his performance in a national competition for the design of the Chinese Art Palace. Later, his repeated success in various other projects have provided him with widespread recognition, leading many who see his designs to remark: “Ah, this is the work of Wang Yi”.

In 1991, Wang Yi was awarded first prize in the youth group category for his work in the Chinese Art Palace design competition. In 1998, Wang Yi received first prize in his competition entry for the design of the China Arch for New Millennium. In the same year, that design was also one of the projects awarded a place in the list of the “Ten Great Designs of Beijing”.

It is also not easy being a professor in a highly prestigious academic institution. In order to teach, Wang Yi faces great pressure on two fronts:

The first, is being a young architect. He often takes on work which includes planning design, construction documents, and site management. He says that, without knowledge and experience in design, construction documents, and site management, he would feel uneasy in his role as an educator. One of his projects slated to come online early this year is the Sichuan Jiuzhaigou International Hotel. Wang Yi resolutely engrossed himself in this project which he calls both challenging as well as academically significant. In 1994, Wang Yi suffered a car accident which left him with a fear of long-distance bus rides into hilly or mountainous country. However, in order to research Sichuanese and Tibetan ethnic architecture, he overcame his fear so as to find inspiration for his designs.

The second source of pressure Wang Yi faces is as a young scholar. Over the years, he has volunteered to assume responsibility for certain scientific research tasks. For instance, his study of Indian architecture was published in *World Architecture* issue 9006 under the title of "Perfume of the World". Taking the opportunity to be a visiting professor at the University of Rome, Wang Yi visited various countries in Europe. With his understanding of both European and Chinese architecture, he visited India where he became more calm and sober. In 1999, after visiting India, Wang Yi finished his second publication concerning Indian architecture (see *World Architecture* issue 9908), wherein he made more accurate assessments and judgments on Indian architecture, building off the work done by others.

Throughout his career, he has been a diverse individual and cannot be confined to a single identity: he is an educator, he is a designer, and he is a scholar. I believe that, in the interaction between these identities, there remains more to be said.

I have said, that the biggest celebration of young architects can be found within design competitions. As someone who has received two awards from national competitions, Wang Yi has his own thoughts on the subject: "architectural design competitions are the chance for architects to express themselves, to showcase their talents, experiment with new thoughts, and discover new viewpoints".

Celebrations, playgrounds, and cradles all embody a similar meaning: they birth and foster the courage needed for new ideas, they provide confidence, perseverance, and inspiration.

What is inspiration? Wang Yi once quoted the sayings of Qian Xuesen: "inspiration is a way of thinking, inspired thinking is thinking in images paralleled by logical thinking. It is currently misunderstood by most people, but it is nevertheless the most creative way of thinking. Once this way of thinking is mastered, it will greatly stimulate creativity".

The two aforementioned award winning projects are the result of logical thinking, thinking in images, as well as inspiration. The ideas and expectations of architects, the public, and traditional philosophers have all become the content of logic and thinking. For instance, Taijitu, Bagua, urban plazas, central axes, the Temple of Heaven, celestial movement, well developed steel structural arch doors, civilization, greatness, etc. are all analysable and discussable. Inspiration and inspired thinking have all contributed to the creation of architectural and artistic concepts, frozen images which are part of the architect himself. It seems as if they can be felt but not explained.

Louis Kahn once spoke at Harvard in the fifties about his design for the Yale University Art Gallery (1951-1953). Kahn said that he had discovered the most central part of his design philosophy when he was about to compose the concept for various departments. In succession, all drawings such as plans, elevations, details, and overall design were done together in one go. Arata Isozaki believes that the central idea is the architect's concept, the "conception" or "birth" of meaning. Can these central ideas, concepts, conceptions and births be understood as the production of inspiration and inspired thinking by architects (the presence of inspiration, resulting in the mastery of a hundred skills as simple as the mastery of one)? I would like to have Wang Yi, taking his two award wining projects, to make a personal explanation about inspiration and inspiration thinking.

Exquisite concepts are not simple, nor are they simply derived, but are rather the derivative of simple concepts collided together to create the infectious quality of multiplicity. What is needed here is inspiration and inspired thinking. Creative power nurtured by inspiration is incomparably huge, though the production of inspiration requires the long-term accumulation of knowledge. Architectural competitions oftentimes serve as a major push to the development of architectural thought, their role in reflecting new ideas and concepts is crucial. The internal logic and external conditions of an architecture provide several possibilities and basic forms, however the overall form of architecture generates a type of transcendence. In addition, the architectural process from conception to realization is an extremely difficult one, in which ingenuity is often difficult to unravel or clarify. This tends to relate heavily to the architect's personal influences, wherein their life experiences associate with their artistic tendencies. Inspiration is a personalized, creative thinking process. It is able to leap and reach logic as well as imagery which would not be accessible in any other way. The completion of successful architectural works depends on this. Imitative content, although containing similar motivations, desires, and knowledge of other artists and architects, lacks the inspiration and inspired thinking which is also central to this process.

As a teacher, Wang Yi possesses a wealth of knowledge, ideas, and experiences to guide his students in logical thinking and imagery, resulting in bursts of inspiration and success. So far, from the works of Wang Yi (including his two award winning projects), this is all that we can conclude.

Originally published in *New Architecture*, 2000 (01)

本土地景 LOCAL LANDMARK

文脉更新 CONTEXT RECONSTRUCTED

本土建构 LOCAL TECTONICS

城市空间 URBAN SPACE

本土再造 LOCAL UPCYCLING

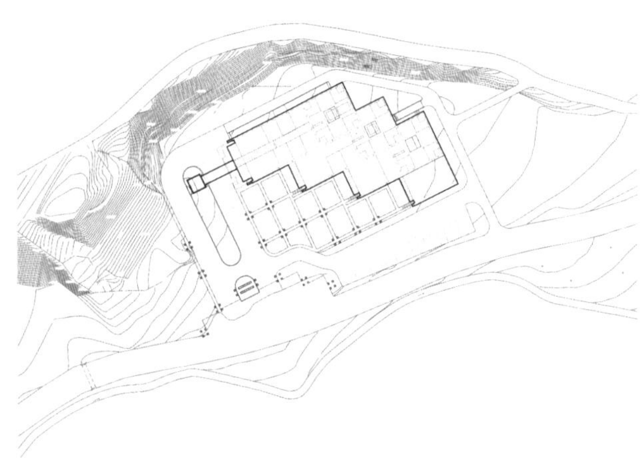

黄龙瑟尔嵯国际大酒店
Huanglong Seercuo International Hotel

札记：从被动到主动的地域建筑

Note: From Passive to Active Regional Architecture

地域建筑是地域文化在物质环境和空间形态上的体现，它体现的不仅仅是建筑的物质形态和使用功能，更是建筑所蕴含的地域文化的深层内涵。全球化所带来的新的建筑材料和技术，以及时尚的设计理念，对地域建筑产生了不同程度的同化。这种同化作用导致了城市特色的没落和建筑形式的雷同。在全球化的语境下，地域建筑的研究逐渐被建筑界所重视，地域性和地域文化成为观察和思考建筑问题的背景和切入点。

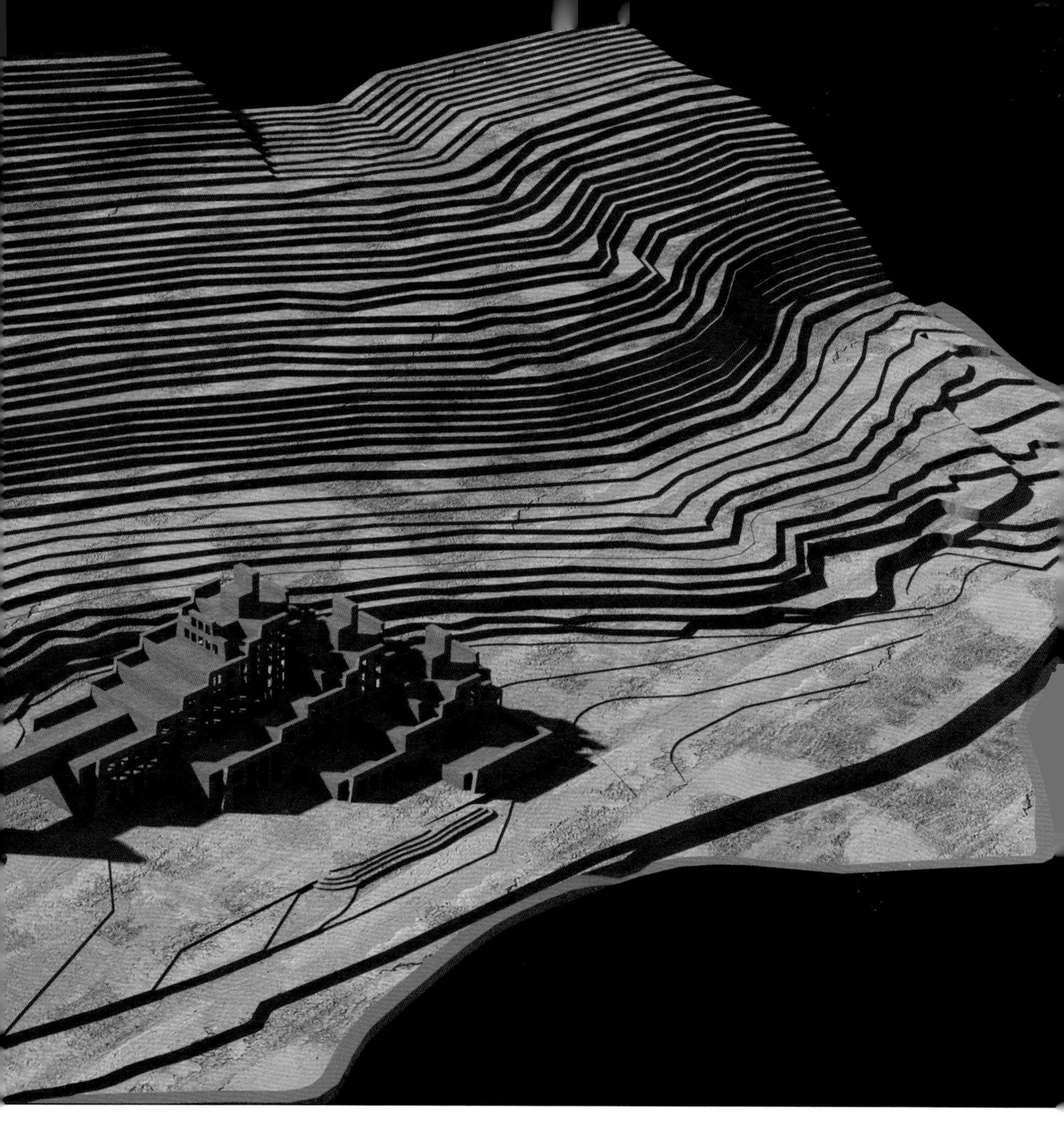

1. 被动的地域建筑

地域主义建筑作为一种建筑思潮开始于 1930 年代，起初它是作为现代主义建筑的抗衡力量出现在世界建筑舞台的。地域建筑首先强调的是尊重地域的地理特征，包括地形、地貌以及气候条件等。地域性是建筑的基本属性，以下将从气候及地理、人文及传统、材料及技术等 3 个方面来概况和理解地域建筑的这些基本属性，以及它们之间的相互影响。

· 气候及地理

自然环境是建筑赖以存在的物质前提，不同的自然环境，有着不同的气候及地理条件，建筑无不是地域气候及地理条件的反映。从“构木为巢”的干阑式木屋到 “挖土为穴”的窑洞式民居，这些具有强烈地域特征的建筑是人们经过千百年来摸索与总结的成果。印度建筑师查尔斯·柯里亚(Charles Correa) 经过研究发现，许多印度传统建筑为适应炎热气候，巧妙地结合了自然条件，他以此参照，

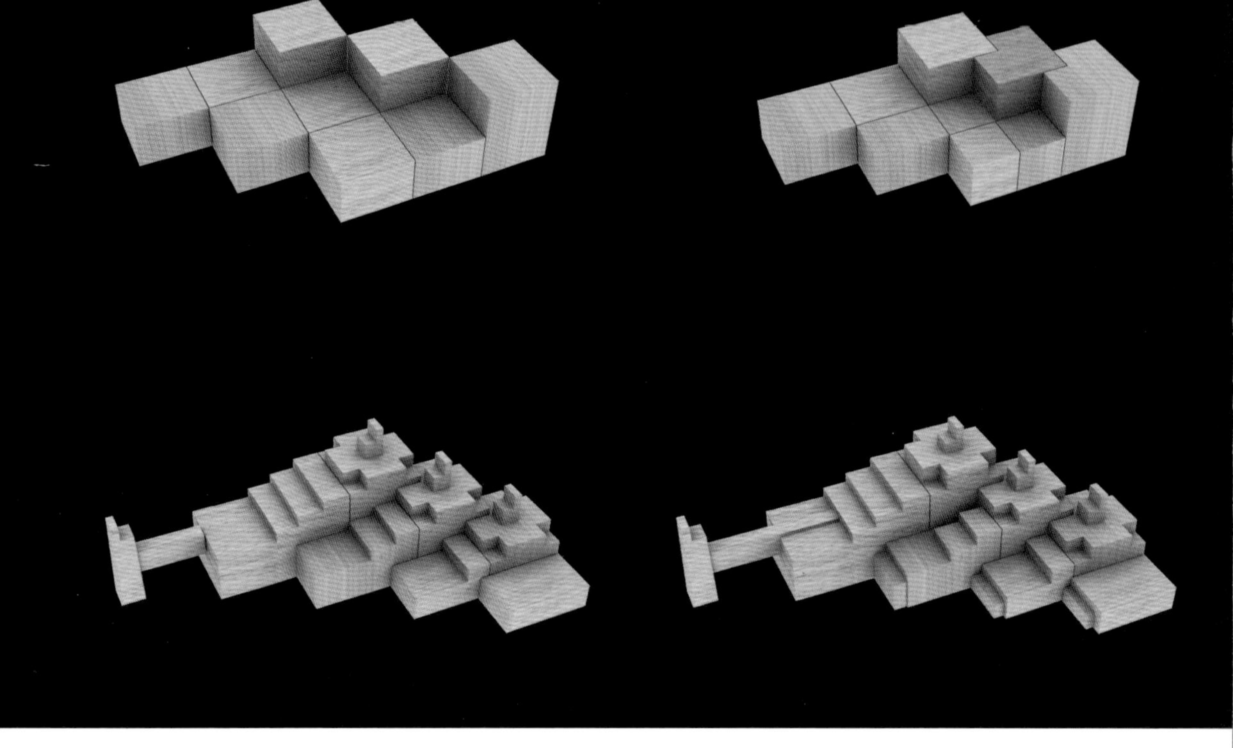

设计了许多具备现代功能又有强烈地域特征的建筑作品。

·人文及传统

地域建筑的第二类属性是人文及传统。人们在与自然共生中形成的传统习俗是长期积淀下来的行为模式，它反映了人们的社会价值取向和需求，反映了地域文化的深层结构，对地域建筑特征的形成起到潜在的影响。如北京的四合院，是反映我国"尊卑有序"的传统居住方式的建筑典型；又如客家族的围屋平面及厚重外墙，是抵御侵袭的最佳居所，也是客家文化在建筑形式上的反映。

·材料及技术

传统地域建筑使用的材料大多取自于自然。地域材料及其建构方式的不同使建筑的表层质感产生差异，地域建筑是地域建筑材料和建造技术普遍发展的一个体现。正如梁思成先生所言："其结构之系统，及形式之派别，乃其材料环境所形成"。如蒙古包是牧民根据条件采用便于拆装的支撑杆件和

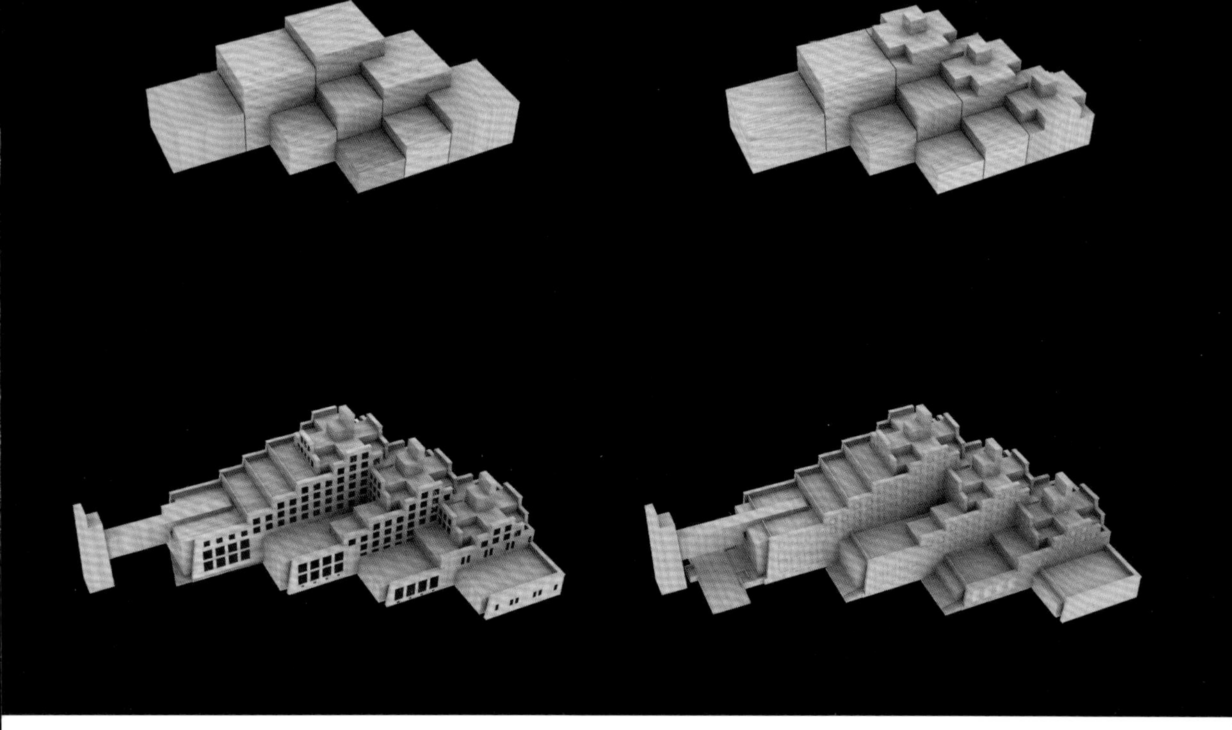

柔软易折的围护毛毡组装的游牧民居。而芬兰的建筑大师阿尔瓦·阿尔托(Alvara Alto)，采用当地的自然材料——木材与红砖，使他的建筑作品具有强烈的地域风格和传统文化特色。

还需指出，以上论述的地域建筑的 3 类基础属性并不是相互孤立的，而是彼此影响、共同对地域建筑的特征发挥作用的。比如，地域材料及技术中同样携带了大量的地域文化信息，同时地域建筑中的人文及传统内涵也无法脱离材料和技术而独立存在。所以在挖掘和保护建筑的地域性特征时，应将以上各种元素加以概况提炼，并融会贯通。埃及建筑师哈桑·法赛因(Hassan Fathy)在其设计中善于将地域元素提炼和融通，其作品被誉为“在东方与西方、高技术与低技术、贫与富、质朴与精巧、城市与乡村、过去与现在之间架起了非凡的桥梁”。

近年来，全球化的推进愈发主动，而地域建筑的调整往往很被动的。建筑的地域属性被全球化所侵蚀，不知不觉中被统一化和同质化。

2. 主动的地域建筑特征

在全球化语境下，如何使地域建筑保持自身的特征而又汲取全球化带来的进步，甚至主动地参与到全球化的进程当中？如何使地域建筑成为现代的而又回归本土的资源，使其呈现出可持续的发展？为此地域建筑必须改变被动调整，而进行主动参与。

· 动态

梁思成先生曾说：“建筑之始，产生于实际需要，受制于自然物理，非着意创制形式，更无所谓派别。”地域建筑反映了人们对自然环境的适应与利用所形成的的居住生活模式，并经历了时间的考验。但新的地域建筑不是传统地域建筑特征的简单复制和拼贴，在借鉴旧有模式的同时，要汲取现代建筑设计理念，结合新的生活方式，利用先进的技术手段，创造出满足现代功能需求的新的地域建筑。只有这样，才能给地域建筑注入时代的活力，使地域建筑始终保持积极的更新（updated）状态，也就是实现动态的发展。

・异化

全球化给建筑领域带来了科学的建构方法和经济的管理模式，但是，地域建筑那些区别于现代建筑的基本属性，也使其自身具备了顽强的生命力，甚至某种合理性。因此，地域建筑应该在被动接受全球化挑战的过程中，以自身的地域属性来丰富和异化全球化带来的影响。事实上，这也为地域建筑调节自身找到了出路，即借用全球化的某些优势，如新的材料和先进的技术等，来调解地域建筑自身与全球化相比所显现的不足。吸收全球化的这些实用性优点，才是地域建筑持续发展的根本途径。

・开放

传统地域建筑产生于自给自足的、相对封闭的系统之中。传统地域建筑所反映的价值观念、社会需求、行为准则、生活方式等，在当代而言，已发生了巨大的变化。地域建筑要要满足当今的社会生活，体现时代的风格特质，就必须具备开放的姿态。正如肯尼思・弗兰姆普敦(Kenneth Frampton)提倡的"与限制的地域主义相反，有另一类型地域主义，就是解放的地域主义"。地域建筑应吸收新技术、新观念、新材料来适应人们现代化的生活方式，反映新时代的特点，实现地域建筑的更高层次的发展，并在对新事物的开放中体现自身存在的价值与意义。

3. 地域建筑设计实践解读

黄龙瑟尔嵯国际大酒店位于四川省阿坝州黄龙国家级风景区沟口，这里从地景、原型和建构等 3 个方面说明建筑设计的理念以及设计手法，从而进一步阐释建筑师在设计中对地域建筑的基本属性的探索和尝试。

・地景

地域建筑的设计首先要体现对地域自然环境的尊重，避免对地景造成破坏。黄龙瑟尔嵯国际大酒店是一座按五星级标准设计建造的旅游接待酒店，位于被收入联合国"世界自然遗产"名录的黄龙风景区。故映衬美丽地景成为酒店建筑设计理念的出

发点。

为了与优美的自然地景相协调，黄龙瑟尔嵯国际大酒店整体形态上借鉴了当地藏羌民居山寨的特色。建筑体量呈阶梯状，前低后高，依山傍势，层层叠落。这种处理手法将建筑各部分化整为零，减弱了建筑物因巨大体量所带来的不利影响。同时建筑顶部3个依次升高的透明玻璃体突出于整个建筑天际线，与周围起伏的山峦渗透呼应。建筑与山地地形和谐而有机，不仅没有破坏原有地景风貌，甚至强化了原有的地形地貌特征。通过将建筑整体形态与地景特质建立起直接或象征性的联系，使人们感受到建筑深深根植于自然环境所获得的某种生命力。

·原型

地域建筑的设计要体现对人文传统的继承，有意识地挖掘地域建筑的原型与意义、地域建筑的空间模式和象征性语言进行设计的再创作。阿坝是藏族羌族自治州，有着浓郁的藏羌文化传统，因此展现地域文化成为黄龙瑟尔嵯国际大酒店设计的另一重要理念。

在建筑整体形态上，酒店设计选用了当地藏羌民居独具特色的碉楼作为母题原型。三栋碉楼有韵律地一字排开，并与一栋独立的碉楼相呼应。母题的反复出现强化了藏羌山寨的空间意境。在建筑墙面上，酒店设计借鉴了藏族多层佛阁建筑外墙的收分做法，使建筑形体呈现出厚重感和雕塑感。在外立面细部上，酒店设计吸收了藏族地域建筑的装饰性元素，如女儿墙、窗套、门头、柱子的做法。但这些传统地域建筑元素进行了中和与简化，变得简洁明快，朴素典雅，以与现代化酒店的身份相适应。

·建构

地域建筑设计既要将现代材料及技术加以创造性地发挥，也要探寻地域材料在现代条件下的合理应用。为此，黄龙瑟尔嵯国际大酒店的外墙面材料采用了现代材料和地域材料相结合的手法。

酒店整体着力体现藏羌山寨的空间意境，但建筑典型特征碉楼因采用了现代材料——玻璃来表达，而散发出一股时代气息。在建筑主体墙面的米黄色采自黄龙风景区内著名景点黄龙梯湖——“金沙铺地”的色彩，同时也是藏族寺庙惯用的颜色；墙顶的藏红色压条也是藏族建筑惯用的做法；部分用铜皮装饰的墙面，则借鉴了藏族寺庙常用金属皮做装饰的传统；建筑底层深色毛石墙面，则是严格采用了当地的石材及其传统砌筑技术。地域材料和做法与现代材料和技术在对比中有机地结合在了一起，创造出具有地域特色又具时代感的建筑形象。

4.结语

全球化既是一种事实，也是一种趋势，它引发的文化趋同，抹煞了建筑的地域特色，导致了大量千篇一律的建筑和城市的出现。但全球化也具有积极的一面，全球化促进了经济发展和技术交流。地域建筑的根本出路在于将自身置入全球化的语境之中，在这个过程中证明自身的价值并获得发展。

传统地域建筑自身封闭的状态制约了它的发

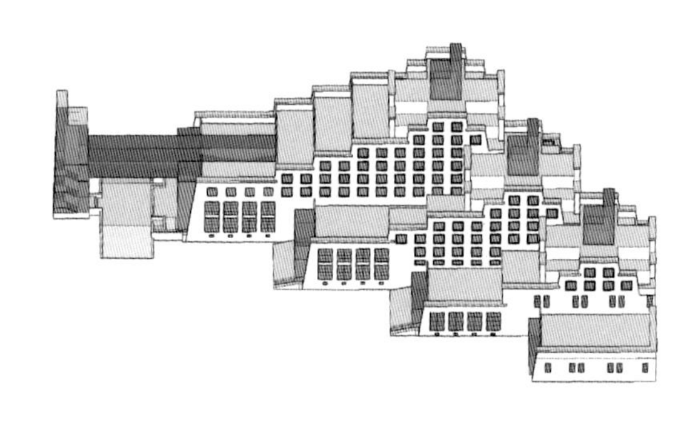

展。地域建筑要重新获得生机，必须进行积极主动地调整和适应。地域建筑不能成为僵死的模式，无论内容与形式、技术与艺术，它的生命力在于要随时代发展。要以开放的姿态，保持自身地域属性的同时，丰富和异化全球化带来的影响。

建筑的地域属性依然是今天代建筑创作的源泉。研究地域建筑，要从产生它的地域环境、生活习俗、技术条件诸多方面来寻找其地域属性的基本规律和内涵。对地域建筑的传统元素进行选择、提炼、抽象和转化，并使之符合时代要求。地域建筑可以提供给建筑师某些设计思路和模式语言，但建筑师难能可贵之处是在此基础上作出某些提升和突破。

原文刊于《建筑学报》2013 年第五期

Regional architecture is the embodiment of regional culture in both its material environment and spatial form. It represents not only the material forms and functions of architecture but also the deep connotations of regional culture contained therein. New building materials, technologies and trending design theories brought about by globalization have assimilated regional identity in various degrees. This assimilation has led to both the decline of city character as well as the duplication of architectural form. In the midst of globalization, study into regional architecture is gaining more and more traction in architectural practice. Simultaneously, regionalism and regional culture serve as the starting point and context for observations and thoughts concerning architectural issues.

1. Passive Regional Architecture

Regional architecture was born as an architectural thought-trend in the 1930s, originally emerging as a contending power against modernist architecture. Regional architecture mainly stresses respect towards regional geological features such as landform, climate conditions, etc. Regionalism is the foundational attribute of architecture. This thesis introduces regional architecture's three basic concerns: climate and geology, humanity and tradition, and material and technology. This work will then explain the effects they have on each other.

· Climate and Geology

A natural environment is the material prerequisite for architecture's existence. Different natural environments are characterized by varied climates and geological conditions, and architecture is the reflection of such. Whether it is the log cabin on stilts embodying "a nest of timber", or the cave dwelling characterizing

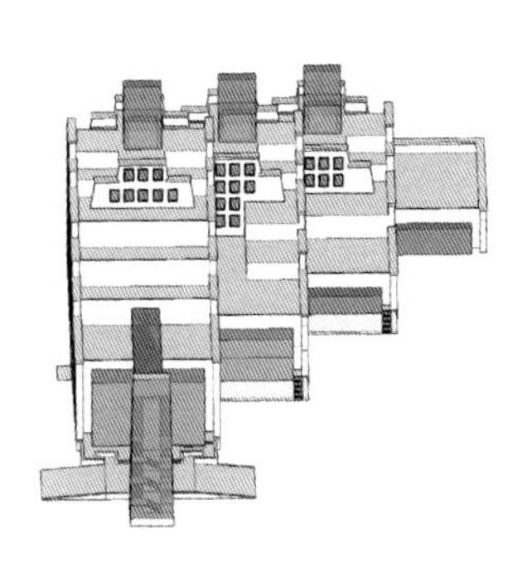

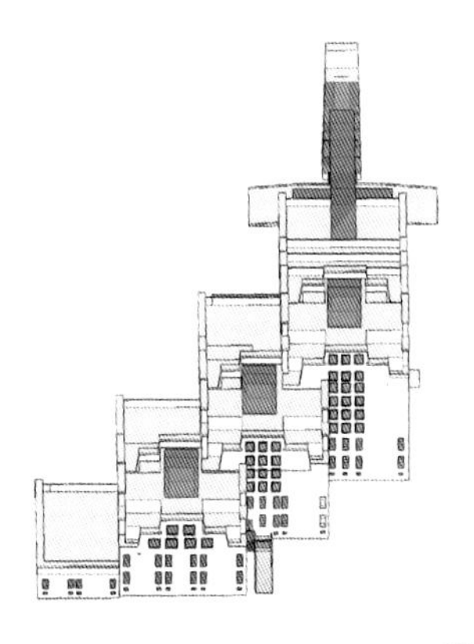

"digging an earthen den", architecture with strong local connotations signifies the built record of a people's experimentation and history spanning thousands of years. For instance, Indian architect Charles Correa has, through his research, found that many traditional Indian structures properly integrate with the natural environment in order to adapt to the hot weather. He has utilized this guideline to create many architectural works with modern functions and strong regional characteristics.

· *Humanity and Tradition*

The second attribute of regional architecture is humanity and tradition. Traditional customs and behaviours accumulated over time are formed through the process of man's coexistence with nature. They reflect the social values and orientations of a people, and signify the

deep structure of regional culture, therefore having a significant potential impact on the formation of regional architectural features. For example, the Beijing courtyard house (*siheyuan* in Chinese) is an architectural typology which adheres to the idea of "emphasize hierarchy and order", reflecting China's traditional mode of life. Another example can be found within the round dwellings of the Hakka people in China: characterized by a circular plan and thick walls, this typology is suitable for defending from invasions and serves as an architectural embodiment of Hakka culture.

· *Material and Technology*

Most of the materials used by traditional regional architecture can be found within nature. The differences in regional materials and their modes of usage result in differences in architectural appearance. Regional architecture includes, in turn, the popular development of these materials and construction technologies. Renown Chinese architectural historian Liang Sicheng once said: "the structure and form of architectural trends are informed by their material environments". For example, the yurt is a kind of nomadic dwelling built by herdsmen using easily transportable structural elements and soft felt, staying true to local conditions. In addition, Alvar Aalto achieves strong regional feeling and cultural characteristics by taking advantage of local natural materials—wood and red brick.

It stands to mention that the above attributes of regional architecture act interdependently instead of in isolation, and they impact the features of regional architecture in concert.

For example,regional materials and technologies contain a large amount of cultural information. Simultaneously, humanity and tradition cannot exist independently without material and technology. Therefore, in the process of discovering and revitalizing regional characteristics of architecture, the above-mentioned elements should be extracted and digested together. Hassan Fathy, an Egyptian architect, is adept at using regional elements in design, as a result his works are often praised as "building a fabulous bridge between east and west, high tech and low tech, poor and rich, simple and complex, city and village, and past and present.

In recent years, the promotion of globalization is becoming increasingly active, while the evolution of regional architecture is oftentimes

passive. Regional attributes of architecture are being overtaken by globalization, resulting in insensitive uniformity and homogeneity.

2. Active Regional Architectural Features

The question stands: how, under the context of globalization, to ensure that regional architecture maintains its own character while utilizing the advancement brought about by globalization (perhaps to the point of even influencing globalization actively)? How to foster a simultaneity between modern and regional resources so as to develop sustainably? Regional architecture must shift from passive appeasement to active participation.

· *Dynamism*

As Liang Sicheng once said:"Architecture begins at practical need and is limited by natural physics. It is not bound by special forms and schools." Regional architecture reflects the way of life formed by adaptation to and utilization of the natural environment, achieved through the test of time. However, new regional architecture should not be mistaken as simply a duplication or imitation of the features of traditional regional architecture. While referencing old patterns, new regional architecture also pulls from modern design theory to address current standards of living. Contemporary technologies are integrated in order to create new instances of regional architecture which satisfies the demand for modern functions. Only in this way can the vitality of regional architecture be recognized in a contemporary state, therefore realizing dynamic development.

· *Alienation*

Globalization has brought about the influx of

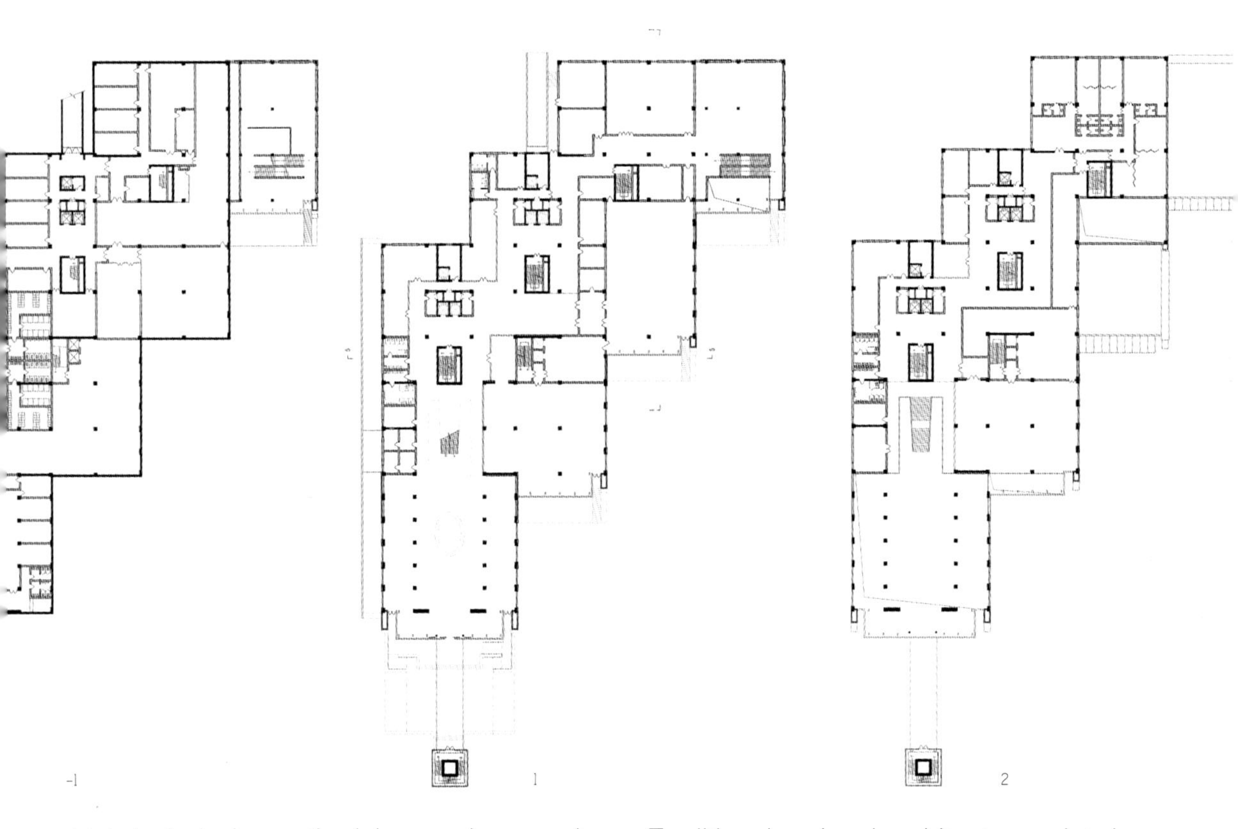

high-tech design methodology and economic considerations to architecture. However, regional architecture differs from modern architecture, it is characterized by strong vitality and unique rationality. Therefore, in order to adequately receive the challenge of globalization, regional architecture should emphasize proprietary regional attributes to address the impact brought about by globalization. Regional architecture should adjust in order to take advantage of certain changes brought about by globalization, such as new materials and advanced technologies, in order to adjust its own realized shortcomings. Regional architecture's sustainable development should appropriate the changes instilled by the advent of globalization.

· *Adaptability*

Traditional regional architecture exists in a self-sufficient and relatively closed system. However, values, social demands, codes of conduct and living styles reflected by traditional regional architecture have changed dramatically to this point. In order to satisfy the needs of contemporary social life and demonstrate the characteristics of the age, regional architecture must adapt an open-minded attitude. The foundations for this exist in the works of Kenneth Frampton, who promotes the dichotomy between limited and liberal regionalism. Regional architecture should absorb new technologies, new concepts and new materials to adapt to a modern lifestyle, reflect characteristics of contemporaneity, realize higher-level development and demonstrate the value and significance

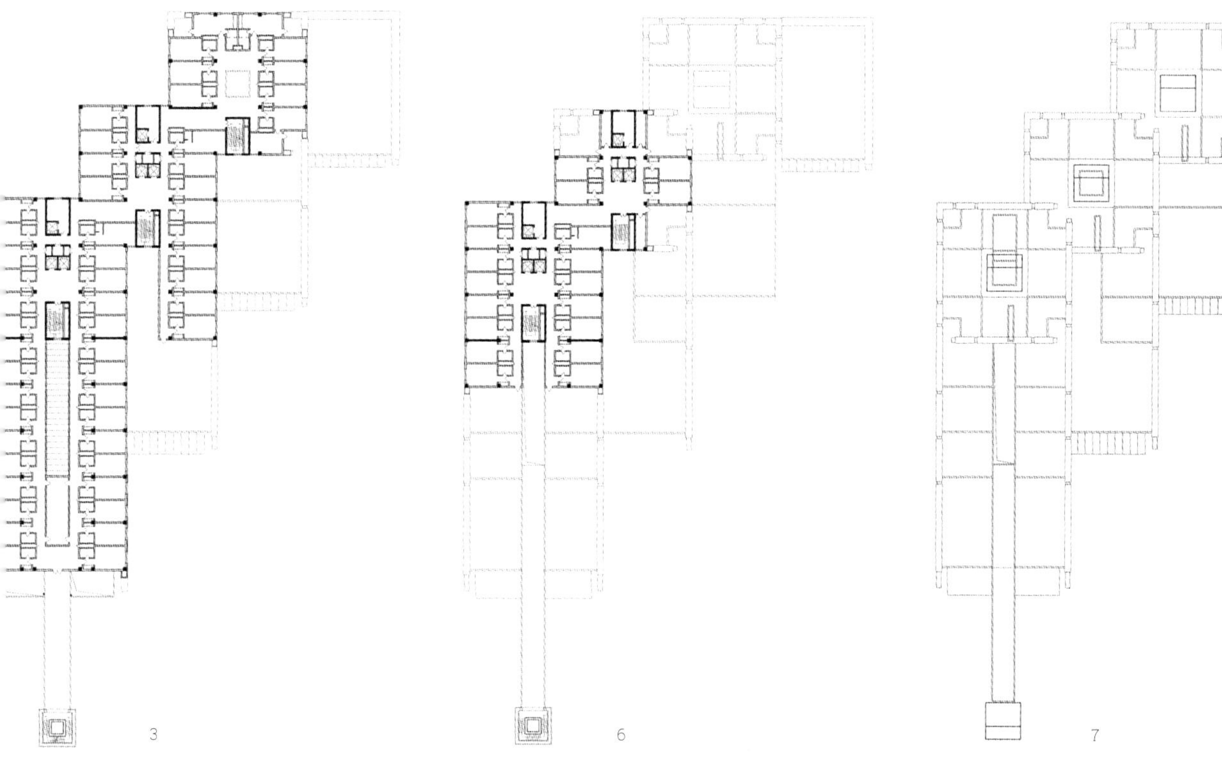

of adaptability.

3. Understanding Regional Architecture's Design Practice

The Huanglong Seercuo International Hotel is located in the Huanglong national scenic area of the Aba Tibetan and Qiang Autonomous Prefecture in Sichuan Province. The below text will explain the hotel's design theory and methodology from three aspects— landscape, prototype and construction, so as to better give an understanding of the architect's explorations and attempts towards regional architectural design.

· Landscape

The design of regional architecture should firstly show respect to the natural environment and avoid destroying the local landscape. The Huanglong Seercuo International Hotel, as a five-star hotel mainly catering for tourists, is located in the Huanglong scenic area, a site listed on the UNESCO's "World Natural Heritage List". Therefore, emphasis of the beautiful local landscape is the starting point for the hotel's design.

In order to coordinate with the beautiful local landscape, the Huanglong Seercuo International Hotel borrows the character of local Tibetan and Qiang residences in its form. The building has a stair-step shape with a low front and high rear, reflecting the rising peaks of the mountains. This method breaks the whole building into sections, easing the negative impact brought on by the considerable size of the project. At the same time, three glass sections rising in turn at the top of the building are higher than the project's overall skyline, echo-

ing with the surrounding rolling mountains. The building coexists harmoniously with the local mountainous landscape, strengthening contextual cues instead of destroying them. Through the building's direct and symbolic connections to the landscape, a sense of vitality is gained by rooting architecture in nature.

· *Prototype*

The design of regional architecture should show the inheritance of a humanist tradition, and consciously sift for local cues towards prototype, meaning, spatial form and language to reutilize in design. Aba is an autonomous prefecture of the Tibetan and Qiang nationalities, possessing a strong sense of Tibetan and Qiang cultural traditions. Therefore, displaying regional culture has become another important design theory considered in the design of the Huanglong Seercuo International Hotel.

The design of the hotel's overall form follows that of the watchtower and incorporates unique characteristics from local Tibetan and Qiang dwellings in its prototype. Three watchtowers line up and echo with an independent watchtower. The repeated appearance of the prototype harkens to the spatial atmosphere of Tibetan and Qiang residences. The design of the hotel's wall borrows construction methods utilized in the outer wall of Tibetan multi-layer pavilions, creating a sculptural sense of profoundness. The details of the outer wall absorb the ornamental elements of Tibetan architecture, utilizing features such as Tibetan style parapets, windows, doors and columns. These elements of traditional regional architecture have been neutralized and refined to be sim-

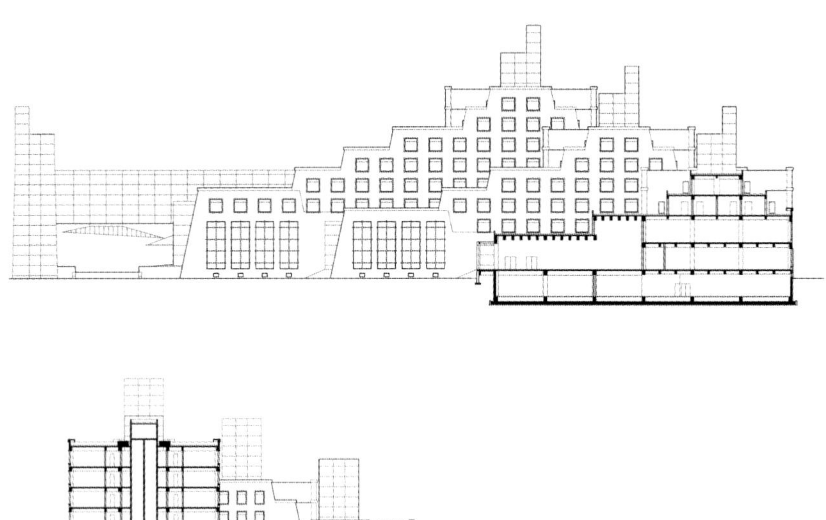

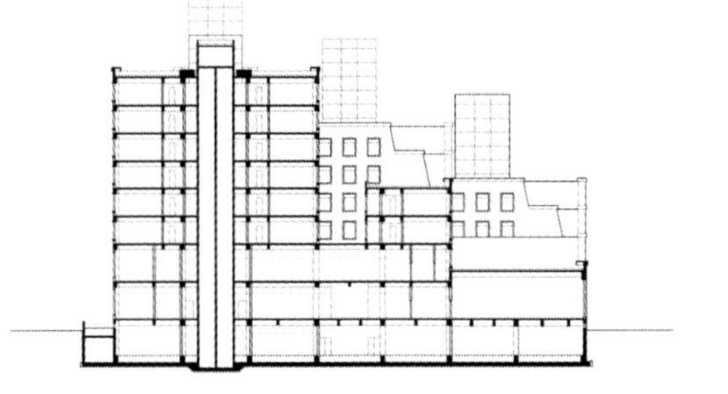

立面图 & 剖面图
Facades & Sections

ple and elegant, so as to suit their inclusion into a modern hotel.

· Construction

The design of regional architecture should creatively provide opportunities to both modern materials as well as modern technologies, exploring the reasonable application of regional materials under modern conditions. Therefore, the material of the hotel's outer wall combines modern and regional materials.

The hotel is intended to display the spatial atmosphere of Tibetan and Qiang residences. The watchtower, which is the prototypical unit of the hotel, utilizes plate glass to endow the building with a modern sense. The beige colour of the walls is influenced by the vista of a famous scenic spot—Ladder Lake in the Huanglong scenic area. This colour is also commonly seen in Tibetan temples. The saffron-coloured batten on the top of the project's walls is also a common element found in Tibetan buildings. The use of copper sheet as a decorative material adopts the Tibetan tradition of using sheet metal as an ornament in temples. The deep-coloured ashlar masonry at the base of the building adopts patterns often seen in local traditional masonry. Regional materials and practices combine organically with modern materials and technologies, creating a building with local characteristics and modern sensibility.

4. Conclusion

Globalization is both a reality and a tendency. It causes cultural assimilation and destroys regional characters of architecture, leading to the emergence of homogeneity in buildings and

cities. But globalization also produces positive effects: it enhances economic development and technological application. Regional architecture should place itself in the context of globalization and prove its own value, adapting itself in this process.

The self-imposed closed system of traditional regional architecture limits its own development. In order to regain vitality, regional architecture must adjust and adapt actively. Whether it is in program or in form, technology or aesthetics, the vitality of regional architecture should develop with the passing of time. Regional architecture should open up to enrich and alienate the impact brought by globalization, whilst simultaneously preserving its own core tenets.

The regional attributes of architecture still

form the foundational guidelines of modern architectural design. Architectural research should study basic rules and patterns regarding subjects such as regional environment, living customs and technologies. Traditional elements of regional architecture should be selected, refined, abstracted and transformed to meet contemporary demands. Regional cues can provide information on existing theories and architectural languages, but the most important role of an architect is to make improvements upon this basic foundation.

Originally published in *Architecture Journal*, 2013 (05)

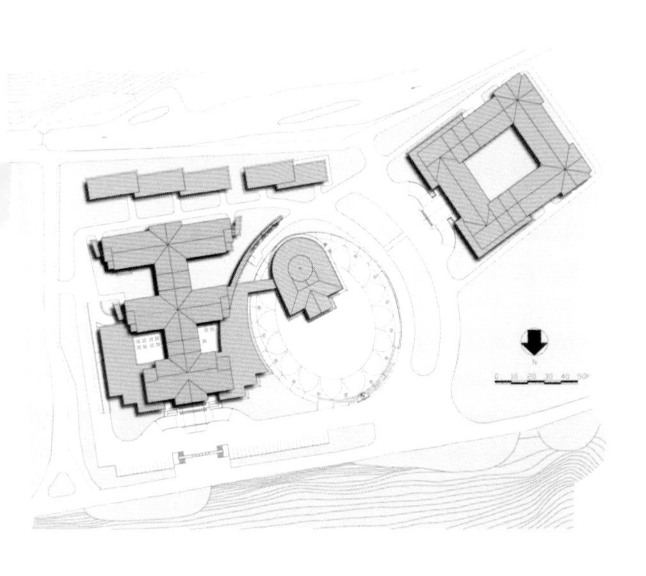

九寨沟喜来登国际大酒店

Jiuzhaigou Sheraton International Hotel

札记：一个结合地域的设计

Note: A Design with Regional Features

· 从狼群说起

动物学家在研究狼群时发现，自然界中的狼群通常都有固定的活动圈，半径大约为 15 公里。有趣的是，狼群的活动圈并不完全封闭，圈和圈之间往往有交叉重叠。也就是说，狼群在划分地盘时，留出了一个共享区域。进一步地研究还发现，活动圈处于交叉状态的群体通常具有较强的繁衍能力，彼此之间的厮杀也少。交叉区域为它们提供了杂交的可能，不交叉区域又使它们保存了各自的特征。

如果活动圈彼此重叠或分离，则会出现要么因重叠而不断厮杀，要么因分离而种族退化。这就是曾在生物学界引起轰动的“交叉圈”理论。

其实，建筑亦有类似“交叉圈”理论的生态特征。在历史的长河中，不同聚居区的人群，根据不同的自然条件和不同的审美意识集体创作出特有的建筑形态，构成自己的建筑 “圈”。在圈内，建筑呈现自己的特征；在圈之间的交叉区，彼此间相互渗透影响，将各自的特征传播出去。这就是建筑的地域性特征。所不同的是，动物的地域性（圈）的形成主要基于种群关系、栖息环境、食物来源等自然因素的影响，而建筑的地域性的形成则要复杂得多，除了受地域气候、地形地貌、环境资源等自然因素的制约外，也深受文化传统、经济发展、社会需求、技术手段等人为因素的影响。

建筑的地域性是建筑的基本特征，也是建筑创作不可忽视的客观存在，只是在过去一个较长的时期内被强大的现代主义建筑所掩盖。在关注可持续

发展的今天，关注建筑的地域性问题更加显得可贵和必要。每个地域都是一个具有特定自然环境和特定文化背景的地方，将建筑放在特定的自然和文化背景之下，才能探讨其存在的意义。脱离了背景，评价就失去了根基。

· 地域的背景

九寨沟位于四川省阿坝藏族羌族自治州，因沟内分布了九个藏族村寨而得名。从地理学角度讲，这里是青藏高原的东南高原亚区向川西平原的过渡地带，其地形地貌复杂，山峦、丘陵、平坝、河流纵横交错。相比青藏高原，这里气候湿润，降水较多；相比川西平原，这里年均温度偏低，日均温差较大。这种处于青藏高原与川西平原的交叉部位的地理气候条件，造就了九寨沟奇特的风景，也孕育出其特有的建筑地域性特征。

刘致平先生在《中国建筑类型及结构》一书中将中国的民居类型分为六大类：穴居类、干阑类、宫室类、碉房类、蒙古包类、舟居类。所谓碉房类建筑主要分布在青藏高原地区，其外墙多以块石或乱石垒砌，因外观像碉堡而得名。这类房屋在历史上早有记载，《旧唐书·吐番传》称："其国都城号为罗些城，屋皆平头，高致数十尺"。屋顶平坦、高墙收分、小窗深凹是这一建筑类型的典型特征。所谓干阑类建筑则分布要广得多，巴蜀是我国古代干阑类建筑的发源地之一。这类房屋底层架空，用来防水防潮，也可防备野兽侵袭。至东汉时期，在干阑式基础之上发展出宫室庭园式建筑，成为四川盆地汉式民居的雏型。而古老的干阑式民居在盆地周边仍然保留下来。唐代诗人杜甫入川时，曾感触巴蜀的这种"殊俗"，写有"好鸟不妄飞，野人半巢居。"的诗句。

位于川西藏东高原亚区的阿坝地区，其民居形态很难简单地归入以上某一类型。这里有地道的碉房式民居，主要分布在靠近青藏高原的山区，也有从干阑式发展而来的，以穿斗架为典型特征的汉式民居，主要分布在靠近川西盆地的岷江下游地区。

在九寨沟一带，由于海拔逐渐降低，气候相对湿润，民居更多地呈现出多样杂交的特征。不同于青藏高原的平屋面藏族民居，这里的民居吸收了汉地民居穿斗架的作法，多做坡屋面，覆盖以石板或瓦片，屋面出挑通常较大，利于排水，以防止夹泥墙或夯土墙遭雨水冲刷。同样由于多山，这里的民居不十分讲究朝向，因势修造，不拘成法，常常在同一住宅中，地平有数个等高。另外，不似汉地民居以平层为主，这里的民居为了节省用地一般为多层为主，即所谓两楼一底式——底层为牛栏猪圈，中层住人，顶层贮藏粮食瓜菜。

·结合地域的设计

“九寨归来不看水”，九寨沟这个被收入世界遗产名录的风景区，有着非同一般的美丽地景。在这样的地景设计一栋建筑是一次非常难得的机会，也是一个不小的挑战。为此，建筑师在如下三个层面上研究了当地传统建筑的地域性特征，并结合到九寨沟国际大酒店的设计当中。第一层面，传统聚落的空间模式；第二层面，传统建筑的造型元素；第三层面，传统建筑的装饰特征。

九寨沟国际大酒店是按五星级标准设计的一座现代化酒店。其规模之巨大、功能之复杂是传统民居无法相比的。处理好建筑与环境在尺度上的协调关系，如何做到化整为零而又零而不散，是设计面临的首要问题。建筑地段位于一段河谷之上，两侧皆为高山。依据这样的环境，参照当地传统村落的空间模式，将整个建筑处理成依山傍势、层层叠落的造型，建筑的各个部分成为既各自独立又彼此相连的整体，颇具一种“山寨”之势。

正如前面提到的，由于地域条件的差异，九寨沟地区的民居有其自身的特点，坡屋顶、穿斗架、夯土墙构成其典型的外在特征。在九寨沟国际大酒店设计中，这些特征被提炼出来作为造型元素，在酒店客房群楼的设计上反复应用。另外，剧场的造型来自藏族的休闲帐篷。这种帐篷通常为白色，在春暖花开时藏族居民外出耍坝子联欢时使用。剧场

前面的歌舞广场采用了莲花瓣的图案，莲花在藏族人心中是吉祥如意的象征。广场上的十二根铜柱顶端动物造型取材于藏族的十二生肖图腾。标志塔的造型则要现代和抽象一些，它融合了佛塔和碉楼的特色，颇具巍峨之势，与主体建筑群体量上取得均衡，成为建筑群竖向构图的中心。

藏族传统建筑的装饰主要体现在女儿墙、窗套、门头、柱子等部位。宗教性建筑装饰复杂，色彩浓重，对比强烈，民居则要朴实简单一些。九寨沟国际大酒店在建筑装饰及色彩方面，吸收了传统建筑的特点并进行了中和和简化，力图简洁明快、朴素典雅，与现代化酒店的性格相吻合，与青山绿水的环境相协调。

看似宽宏的自然环境，承受力是有限的，而且非常脆弱。建筑的价值在于能够被地域环境接纳认同，同时又为其赋予一定的新意。九寨沟国际大酒店作为一个多层次的分散又集中的建筑群综合体，把如此功能不同、规模不同的空间有机地组织在了

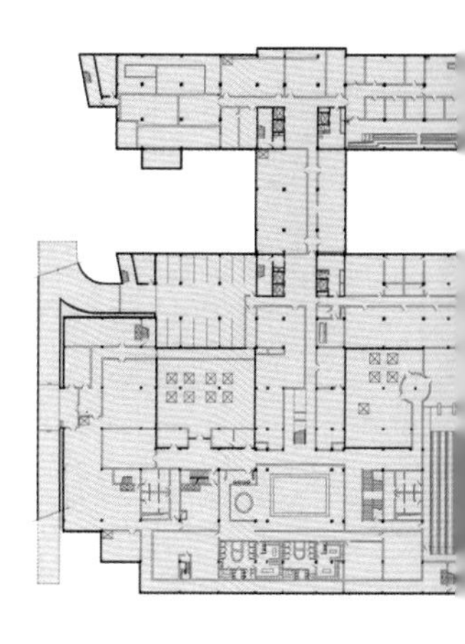

-1

一起，形成了一个新的地域环境认知图标。但整个建筑群并不显突兀，在体量上与环境尺度是协调的，在形态上延续了原有环境的肌理。建筑群的空间是自由的、开放的。新的功能、新的场所，为新的地域活动创造了条件，也为新的地域价值奠定了基础。

·材料和技术的发展

走进藏族村寨，淳朴憨厚的木屋给人恰似归乡的情怀，高耸挺拔的石墙又将人带往久远的过去。材料是有生命和语言的。不同地域的人们根据不同的气候特点和自然环境，经过长期探索，寻找出与之相适应的建筑材料以及营造技艺。藏族人高超的石墙砌筑技术，就是结合当地泥石材料的特性和墙体收分的技巧总结出来的，在几百年前就建造出高度近 50 米的气势恢弘的碉楼。

不可否认，地域性的建筑材料和建造技术对建筑的地域性特征有着非常重大的影响。在工业不发达的过去，天然材料的运用构成了地域性建筑的主要特征。随着技术的不断进步和地区间经济交流的

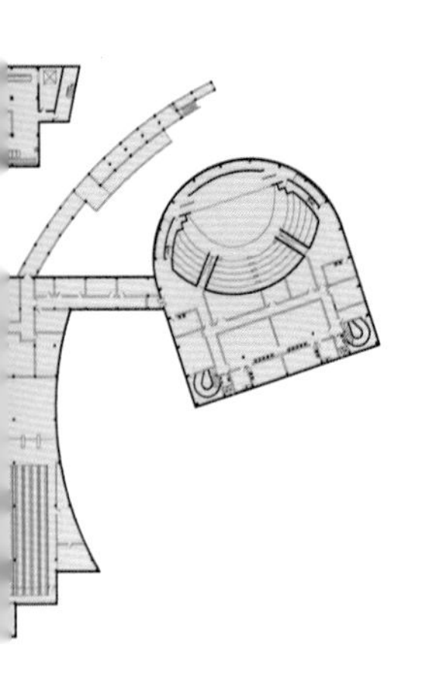
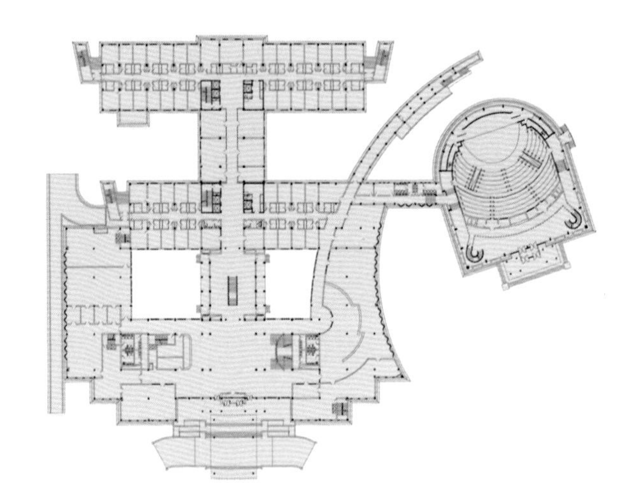
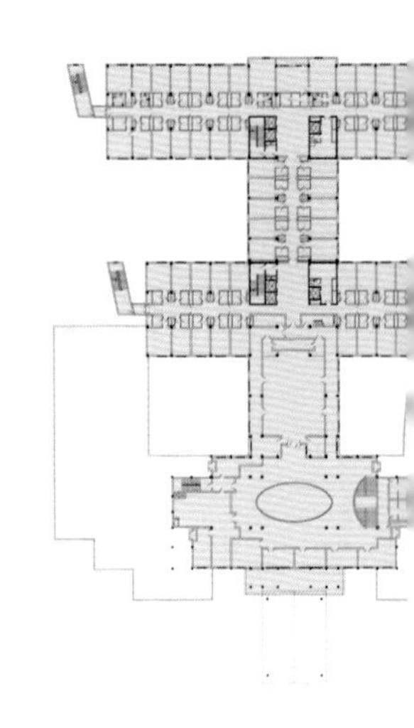

日益频繁，情况已有所改变。人们不再仅仅拘泥于传统的地域性材料和技术，而有机会选择从性能到价格都更优异的新技术和新材料。而另一方面，从现代的生态和环保的观念来看，某些传统的地域性材料对环境也是不利的。如传统建筑中大量使用木材的作法，在森林资源日益贫乏的今天，即使在九寨沟也是不适宜的。又如传统建筑中的石板瓦，尽管很有特点，很有美学价值，其技术性能无论如何也无法与现代的瓦相比，其成本优势在今天也消失殆尽了。一些天然材料和传统工艺在今天已变得昂贵和奢侈，所以一味地强调地方材料的应用并不合时宜，在实践中也往往碰壁。

将大路货的现代材料和现代技术运用于地域性建筑的建造是一个必然的发展趋势。问题的关键在于如何应用。九寨沟国际大酒店将奥地利的灰绿色沥青瓦和德国的白色涂料用在外装修上，有了更好的物理性能和与自然协调的色彩。在空调系统中取消了制冷机组，取山上溪流里温度较低的水来制冷降温。将现代技术与地域条件巧妙结合，以降低能源消耗。

建筑是生活的容器，它的空间形态和营建模式，是特定生活方式和价值观念所生就的。它应该反应民众的需求，而非建筑师的需求。地域性建筑的研究者和设计者，不能仅仅把它看成一门学科，习惯于去考证、去研究，而忽视了建筑首先是生活。无视生活的变化，研究和生活脱离，研究和创作脱离，当学者们走下课堂、走出电视，回到乡村后，发现生活已经变了模样。

·地域性和全球性——有时仅仅一步之遥

从某种角度来看，地域性建筑确实曾是一种类似活化石的东西，它长期顽强地保留着民众古老的生活形态和意识轨迹。就像费孝通所说的："乡土社会在地方性的限制下成了生于斯、死于斯地社会。常态的生活是终老是乡。"然而，这种常态的生活随着全球化的浪潮正在被打破。

虚拟、共享、一体化，人们在分享着先进的技

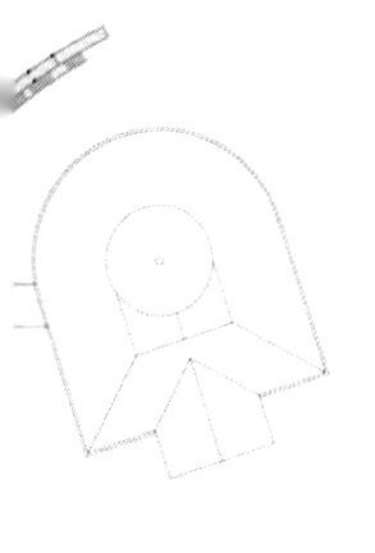
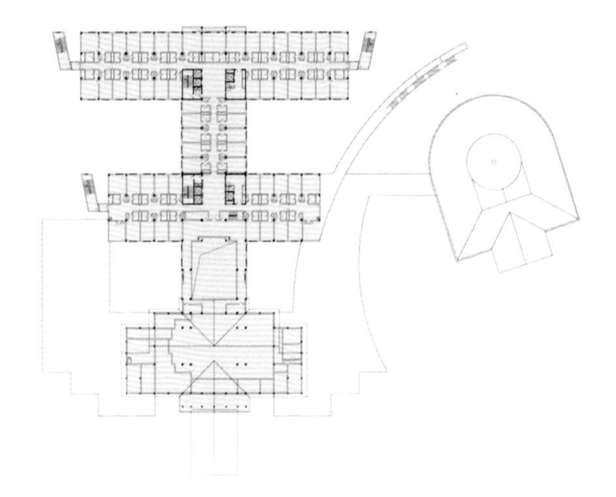
3

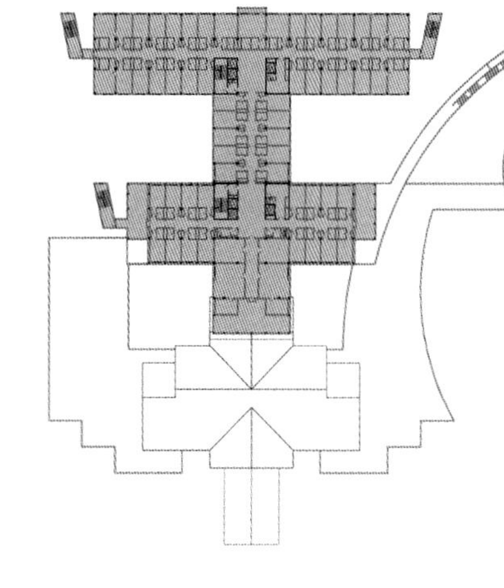
4

术和经验，交流着彼此的信息和产品的同时，地域文化渐渐被遗忘或不屑一顾。在建筑方面导致的结果是，一方面，建筑变得更为高效和舒适；另一方面，建筑形态不断趋同、城市特色千篇一律。

动物学家发现的狼群交叉圈理论，在研究古希腊、古罗马的城邦制度时发现了类似的社会生态现象。古希腊人和古罗马人用圆代表城邦，用相交的圆表示城邦间的融合。在他们的理念中，城邦既不能彼此隔绝，也不能完全融合。只有处于既融合又隔绝的状态时，城邦才是最有生机的。他们把交叉的圆作为图腾刻在了神庙上。后来这个“交叉圈”理论还被借鉴到许多领域，大到解决国际边界争端，小到夫妻相处艺术。

地域性和全球性近几年越来越成为人们关注的世界性话题，建筑界亦不在例外。既要保留地域性特征，又要跟上全球性发展，这是看似矛盾，其实又不矛盾的课题，狼群的交叉圈应该能够给我们不少启示。有人甚至发明了一个新词——全球地方建筑（globl+local=glocal architecture）来个一言以蔽之，想必关注点就在于那个不大不小的交叉领域。

2001 年亚太经合会议在上海举办，与会各国各地政要不分男女老少、高矮胖瘦、黑黑白白，统统都穿上了据说是按西式方法裁剪缝制的中式唐装。一时间这种典型的中式对襟褡袄风靡海内外，连最前卫的好莱坞明星也未免落入“俗”套。不禁让人徒生感慨，信息化的时代，地域性与全球性有时可能仅仅一步之遥。

原文刊于《建筑学报》2004 年第六期

· *Starting with story of wolves*

In their research of wolves, zoologists find that wolves of the same tribe have their own designated circle of activity whose radius is about 15 kilometres. Interestingly, this circle is not a closed one, but overlaps with other circles. Thus, when these tribes of wolves divide their

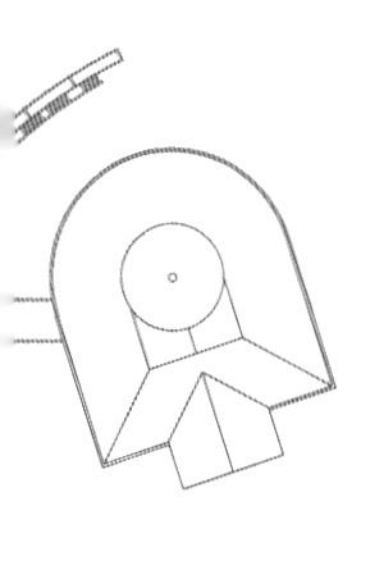
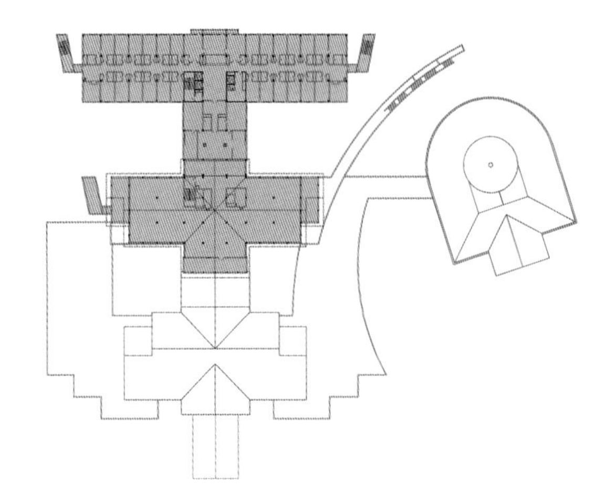
5

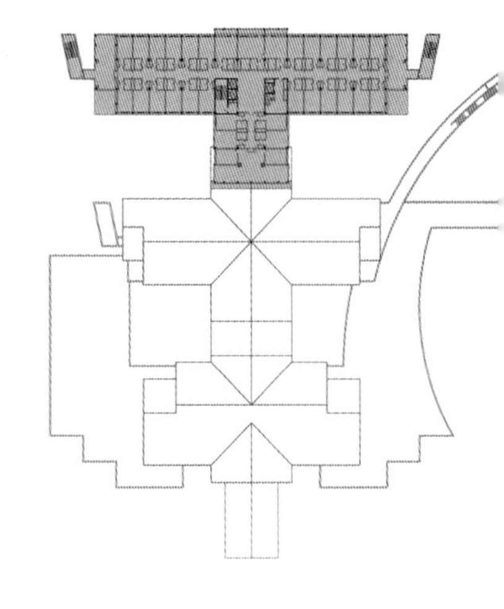
6

territory, they include shared areas. Further research reveals that wolves living in shared areas have better reproductive ability and fight less with members of other tribes. Such overlapping areas provide cross-breeding possibilities while non-overlapping areas help maintain their own characteristics. However, there exists a fine balance, because if the circles overlap completely, or exist totally isolated from one another, fights would spark, either as a result of the integration of territories, or because of racial degeneration as a result of the separation. This is the theory of overlapping circles which has caused quite a stir in the field of biology.

In fact, architecture bears similar ecological features could be explained by the same theory. In the long course of history, people from different habitats collectively created distinct architectural forms based on different natural conditions and aesthetics, constituting their own architectural circles. Inside these circles, buildings display their own features; in overlapping areas of these circles, one spreads its own features and influences another. This is the regionalism of architecture. The difference is that animals form circles on the basis of such natural factors as breed, habitat, source of food, etc, while the regionalism of architecture is more complicated. Except for natural conditions such as climate, geography, natural resources, it heavily depends on human factors including cultural tradition, economy, social demands, technology, etc.

Regionalism as the fundamental building feature is a reality not to be ignored in designing architecture, although it has been blurred

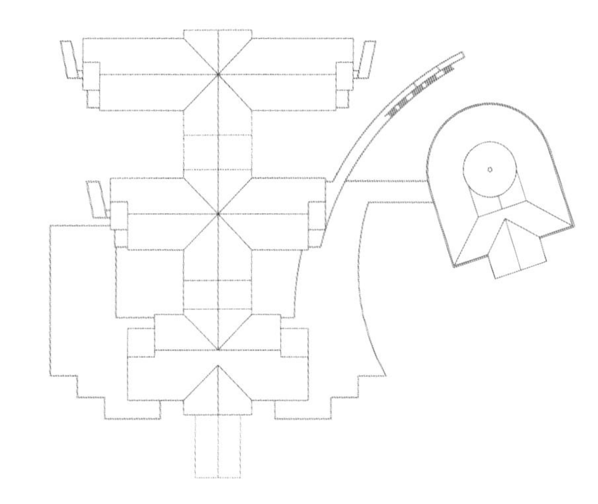

7

for a long period of time by an overwhelming wave of modernist ideologies On the other hand, as nowadays sustainable development is arising it becomes more desirable and necessary to bring the issue of regionalism in architecture to the foreground once again. Every region has its distinct natural conditions and cultural background. Only by putting a piece of architecture into its natural and cultural background we can talk about its significance. Judgments would be baseless without considering the background.

· *Regional background*

Jiuzhaigou Valley (literally "Nine Village Valley") lies in Aba Tibetan and Qiang Autonomous Prefecture in Sichuan province. It takes its name from the nine Tibetan villages along its length. Geographically, this area is a transitional zone stretching from the subregion of the southeast Tibetan Plateau to the Chuanxi Plain. Its complicated landforms, which include mountains, hills, plains and rivers, intersect with each other. The valley has a higher degree of humidity and a larger amount of rainfall than the Tibetan Plateau; the average annual temperature is lower than the Chuanxi Plain and the average temperature difference between day and night is larger. The geographical and bio-climatic conditions that this overlapping zone provides have not only generated the fantastic scenery of the Jiuzhaigou Valley, but also gave birth to the distinct regional features of its architecture.

In The Categorization and Structure of Chinese Architecture, Liu Zhiping categorizes Chinese dwellings into six major types: caves,

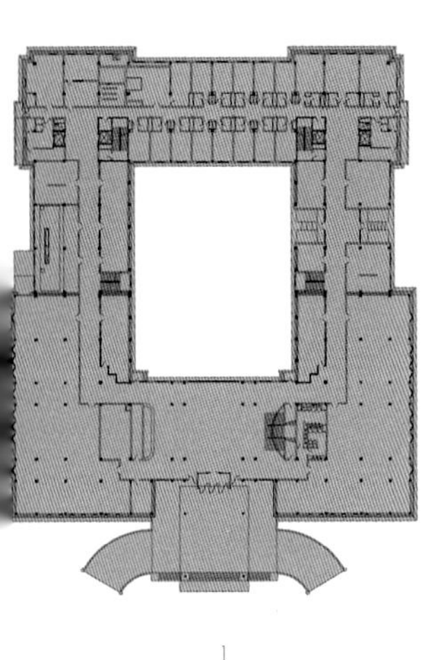
1

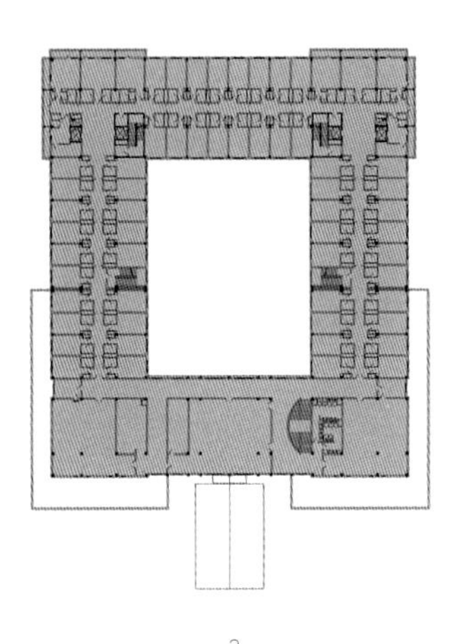
2

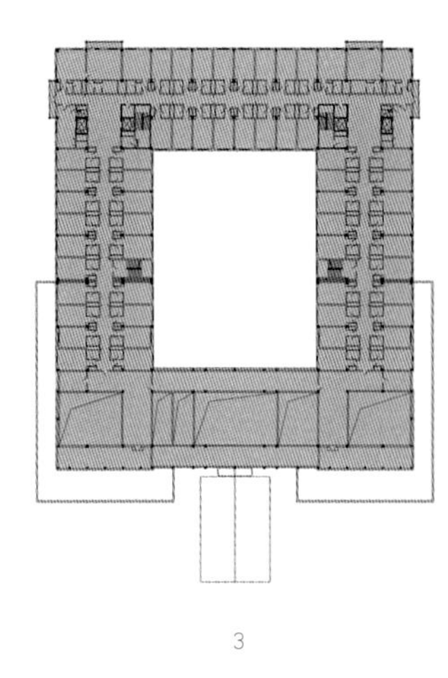
3

stilted wooden houses, palaces, blockhouse-style houses, Mongolian yurts and boat-style houses. In the Tibetan Plateau region the dominant dwelling type is the 'block house'. Their walls are built of rocks and rip-raps, resembling block houses. Such type of building were already recorded early in history, for instance in a chapter devoted to the Tubo empire of the Tibetan Plateau in Jiu Tang Shu (Old History of Tang Dynasty) that says, "the capital of the empire is named Luoxie City,where house measuring ten chi in height with flat roofs were built." Flat roofs, contracted high walls and hollow windows are characteristic of this type of building. Stilted wooden houses, originated from Sichuan province, are distributed more widely in China. In this type, the ground floor is elevated so as to avoid humidity and to fend off wild animals. In the Eastern Han Dynasty, palaces and courtyards were developed on the basis of stilted wooden houses, which then became the prototype of Han-style dwellings in the Sichuan Basin. However, there were some stilted wooden houses remaining on the edge of the basin. When Du Fu, the Tang Dynasty poet, was travelling in Sichuan, he noticed this special custom of the locals and wrote, "birds here do not flutter rashly;half of the wild men nest high in trees".

The dwellings of the Aba region, which lies in western Sichuan and the subregion of the eastern Tibetan Plateau, can not be readily categorized into one particular type. There are genuine blockhouse-style dwellings mainly in mountainous areas adjoining Tibetan Plateau, as well as typical Han-style dwellings

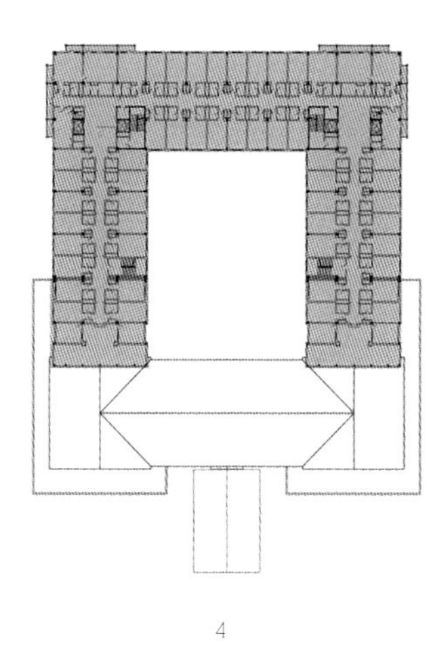

4

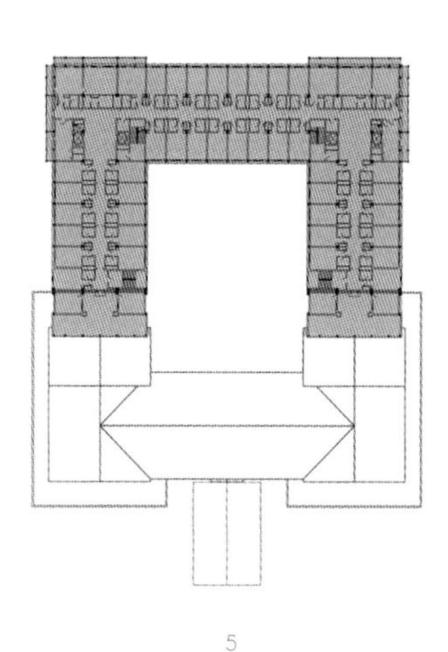

5

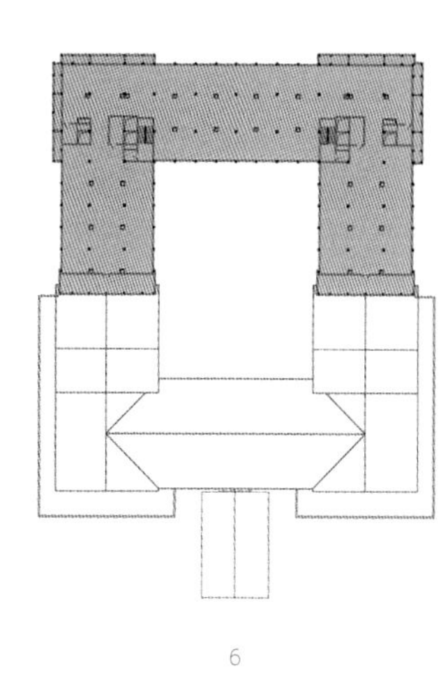

6

with wooden structures, which evolved from the stilted wooden houses, that are distributed mainly in the downstream region of the Min River near the Chuanxi Basin. Along the Jiuzhaigou Valley, as the altitude gradually lowers and the climate is relatively humid, the dwellings take on a more hybrid appearance. Different from the flat-roofed Tibetan dwellings of the Tibetan Plateau, dwellings here mostly have sloped roofs, covered with slates or tiles. Cornices are left with a big margin for drainage and for preventing rain from ruining mud and rammed earth walls. These dwellings, due to their mountainous geographical background, are not particular focused on their solar orientation, but are more built in accordance with the topography. Usually, there are a number of contours in a single dwelling. Besides, unlike Han-style one-floor dwellings, here multiple stories are built to save land, meaning two stories above the ground floor. The ground floor is used as cattle pens and pigsties, the middle story for accommodating people, and the third story for storing food.

· *A design integrating regional features*

It is said that when you come back from Jiuzhaigou Valley, the waters elsewhere would not be attractive any more. Inscribed as a World Heritage Site, Jiuzhaigou boasts magnificent sceneries, making the opportunity of designing a building here extraordinary and challenging. Therefore, the architect looks into the regionalism of traditional local architecture in three aspects and integrates the findings into the designing of the Jiuzhaigou Valley International Hotel. These aspects are: first of all, the spa-

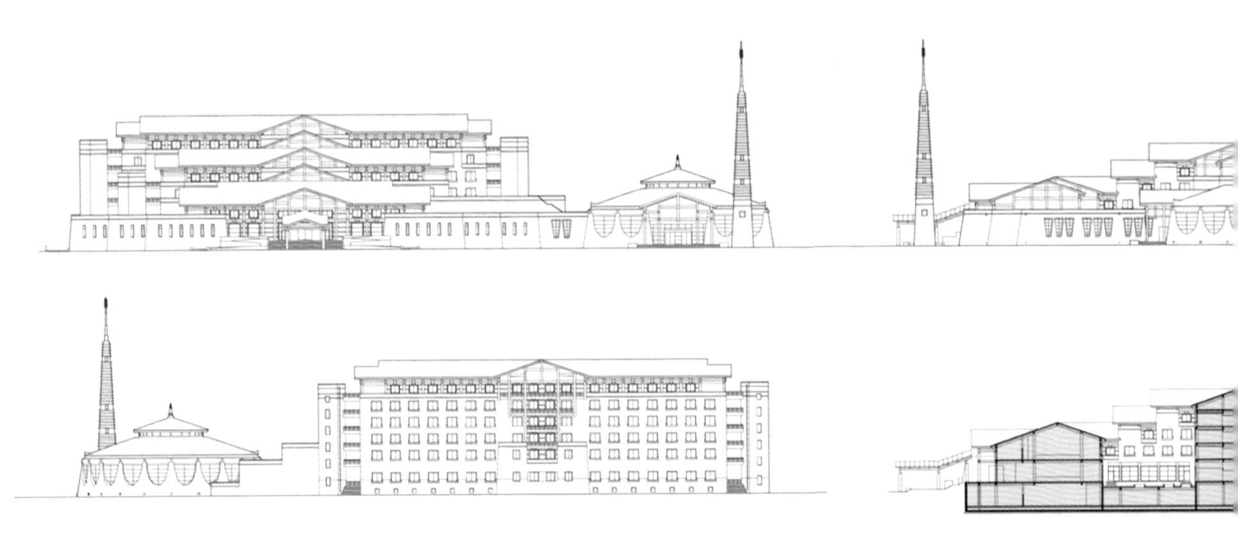

tial patterns of local dwellings; secondly, the modelling elements of traditional architecture and thirdly, the decorative features of traditional architecture.

The Jiuzhaigou Valley International Hotel is designed following a modern five-star hotel standard. Its incomparably huge scale and complex functions are not present in traditional dwellings. How to coordinate the building and the environment is the foremost problem the architect faces. The construction spot is above a section of the valley and on both sides are tall mountains. In view of the geography and the spatial pattern of the local villages, the architecture is kept in accordance with the curves of the topography with one floor stacked upon another. Each part is both independent from and interconnected with each other, making it look like a stacked village.

As I have mentioned, due to regional differences, the dwellings in Jiuzhaigou Valley have their own characteristics, featuring sloping roofs, a wooden structure and rammed earth exterior walls. These features are adopted as modelling elements and repeatedly applied in the guest room design of the hotel. In addition, the theatre is modelled upon tents that are traditionally used for the entertainment of the Tibetan people. These tents are usually white and used when Tibetans go out for fun in the spring. The square in front of the theatre is decorated with a lotus petal patterns—the lotus being the symbol of good luck in the minds of the Tibetans. The animal models on top of the twelve copper columns in the square come from the twelve animal totems of the

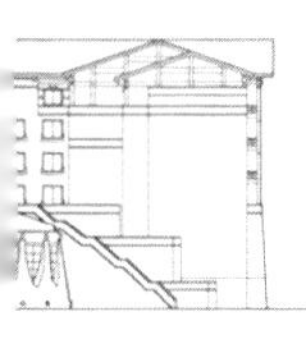
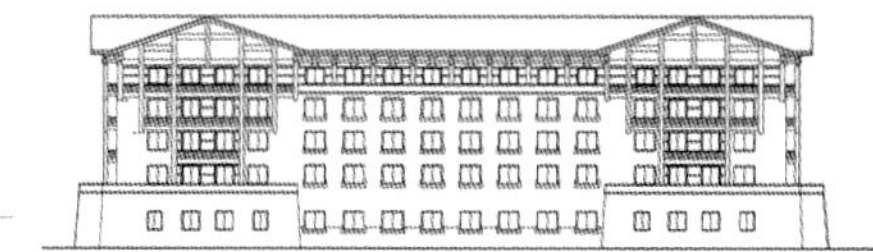

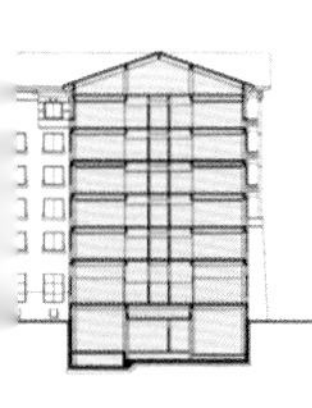
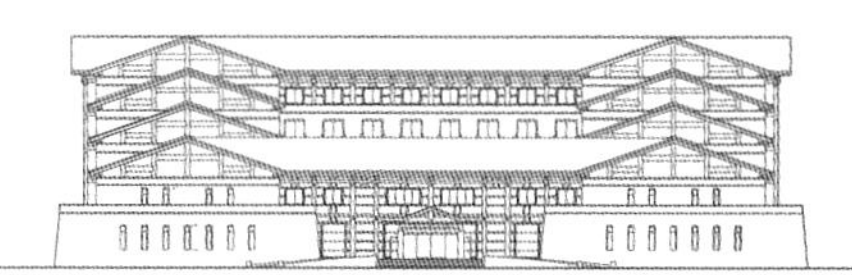

立面图 & 剖面图
Facades & Sections

Tibetan people. The landmark tower, integrating the distinctive features of the pagoda and the blockhouse-style dwellings, looks modern and abstract. Its imposing manner, striking a balance in size with the main complex of buildings, becomes the central point.

The traditional architecture of the Tibetans figures most prominently in places such as parapets,exterior window trims, door heads and columns. Compared with the complexity of decoration and rich colours of religious architecture, folk dwellings are very plain and simple. In aspects of decoration and colour, the Jiuzhaigou Valley International Hotel, while absorbing features of traditional architecture, mitigates and simplifies those features so as to reach an effect of conciseness,liveliness and gracefulness. This is not only in keeping with the character of a modern hotel,but also with the lovely natural environment.

The natural environment, though expansive, has very limited and fragile carrying capacity. The value of a building lies in being adopted and acknowledged by the regional environment and meanwhile bringing novelty to the region. Jiuzhaigou International Hotel, as a multi-layered complex, unites spaces of different functions and different sizes organically and forms a new icon of the region. Nevertheless, the complex does not stand out too much. Its size is coordinated with the environment, and its shape matches the original landform of the place. The spaces of the complex are flexible and open. New functions of the space create possibilities for new regional activities and lay the foundation for

new regional worth.

· *Development of the material and technology*

In Tibetan villages, the plain and simple cottages make people feel as if they were back in their own hometown. The towering stone walls, on the other hand, take people to the remote past. Materials have their own life and language. People of different regions, in view of different climates and natural conditions and through trial and error, have found building materials and techniques suitable to their environment. For example, the Tibetans' superb technology of building stonewalls has taken shape by considering the distinct features of the local mudstones and techniques of building contracted walls. That's how a nearly 50-metres watchtower was erected a few hundred years ago.

Undeniably, building materials and technology of a region have a great influence on the regionalism of architecture. In the industrially undeveloped past, the application of natural material is the prominent feature of regional architecture. Things have changed with the advancement of technology and economic exchanges among regions. People no longer restrict themselves to traditional regional materials and technology, but are able to choose new materials and technology with better performance and price. On the other hand, considering the ecological and environmental consequences, some of the traditional materials are now harmful to the environment. For instance, because the shortage of wood resources has become a pressing issue, the tradition of using large amounts of wood in buildings is not desirable, even in Jiuzhaigou Valley. Slates, though special and aesthetically valuable, cannot match modern tiles in performance. Their advantage of costs is nonexistent either, as natural materials and traditional technology have become much rarer and higher in price nowadays. Therefore, it is not necessarily appropriate to emphasize the use of regional materials and it will cause difficulties in practice too.

It will be an inevitable trend to put mass-produced modern materials and modern technology into the creation of regional architecture. The key of the question lies in how to apply them. For the hotel's exterior decoration, Austrian greyish green asphalt shingles and German white paint is used, whose physical performance are better and whose colours

coordinate with nature. The air-conditioning system has done without the unit cooler and uses cold stream water instead. Such integrations of modern technology and regional conditions can reduce fuel consumption.

Architecture is the container of life. Its spatial patterns and building models are the outcome of particular ways of life and values. It should meet the needs of people, not of the architect. Researchers and designers of regional architecture should not regard it as a discipline for which only research is to be conducted, but should be aware that architecture is part of life in the first place. Any research that breaks away from life's changes and life itself will be futile. When the scholars go back to the countryside from their university and media, they would find that life has changed

tremendously.

· *Regionalism and globalism: sometimes one step is distance*

To some degree, regional architecture used to be living fossils where the folks' age-old ways of life and ideology are obstinately preserved. As what Fei Xiaotong has said, a rural society constrained by its regionalism is where people were born and die, and the standard is for one to spend his whole life there. However, this standard is being broken by the wave of globalization.

Virtuality, sharing and integration are means through which people share advanced technology and experience, exchange with other's information and products. But at the same time, regional culture is being forgotten or disregarded. The consequence upon architecture is that on the one hand, buildings become more efficient and comfortable; while on the other hand, their patterns are becoming uniform and featureless.

The theory of wolves' overlapping circles discovered by zoologists finds similar manifestation in the research of the social ecologies of the city-state systems of the ancient Greeks and Romans were. They used circles to signal city states and overlapping circles to signal the integration of city states. In their ideology, city states could neither be isolated from each other nor be integrated completely with on another. The state of being partly isolated and partly integrated could produce the ultimate vitality. They inscribed overlapping circles to temples as totem. Later, this theory of overlapping circles was borrowed and applied to

many other areas ranging from large scales, such as international border conflicts,to smaller areas such as conjugal relationships.

Regionalism and globalism in recent years have become a topic of global concern,even in the field of architecture. The seemingly contradictory subject of preserving regionalism and catching up with the global development is in fact not contradictory at all, since the theory of wolves' overlapping circles can give us much inspiration. Someone has even coined a new term to capture this trend—"global+local=glocal architecture". In my opinion, the focus must be the overlapping area which is moderate in size.

The 2001 APEC meeting held in Shanghai witnessed a wave of wearing Tang suits tailored in western methods by political leaders,

men or women, young or old, tall or short, thin or stout, white or black, from all over the world. For some time, this typically Chinese traditional clothing was so popular worldwide that even Hollywood stars wore it. It dawns on me that in the era of information technology, there may only be one step's distance between regionalism and globalism.

Originally published in *Architecture Journal*, 2004 (06)

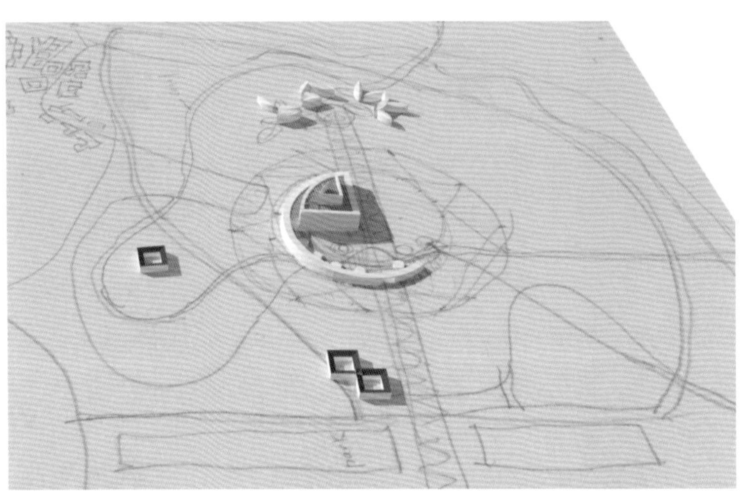

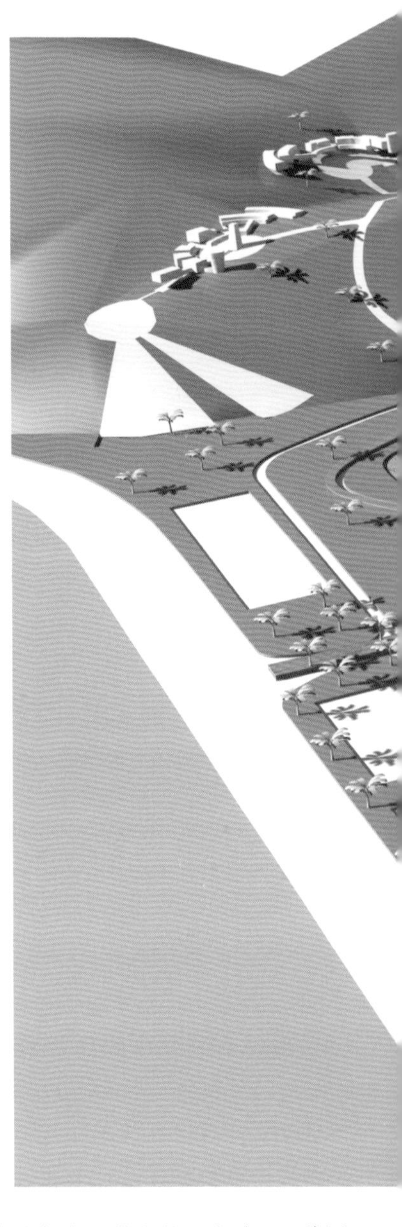

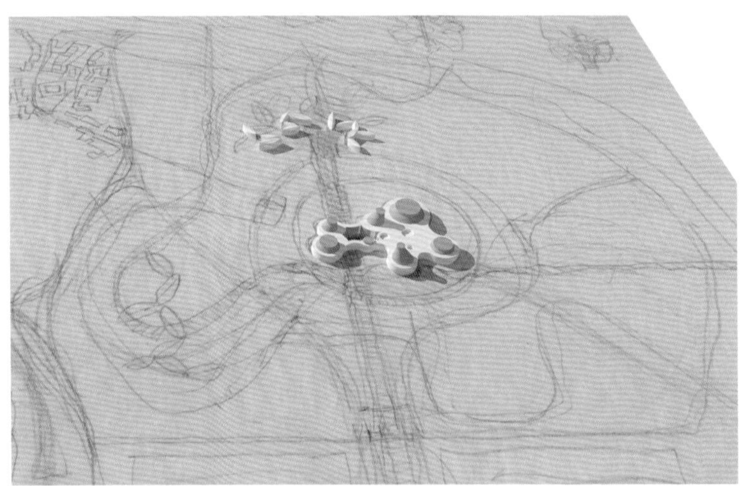

呼伦贝尔民族文化园

Hulun Buir Cultural Park of Nationalities

主题：自然蓝本

Theme: Natural Prototype

该项目位于海拉尔市正北方位，占地100公顷。园区分为15个板块，主要包括展示呼伦贝尔地区少数民族——如蒙古族、鄂温克、鄂伦春、达斡尔、俄罗斯等少数民族的民俗体验板块，滑雪、射箭、跑马等旅游休闲板块，以及位于园区中心的多功能综合体。

园区总体布局吸纳了草原上河流的形态，自然流淌，最大限度地减少人工痕迹，与真正的原生态景观相匹配。综合体由演艺、餐饮、会议、客房、

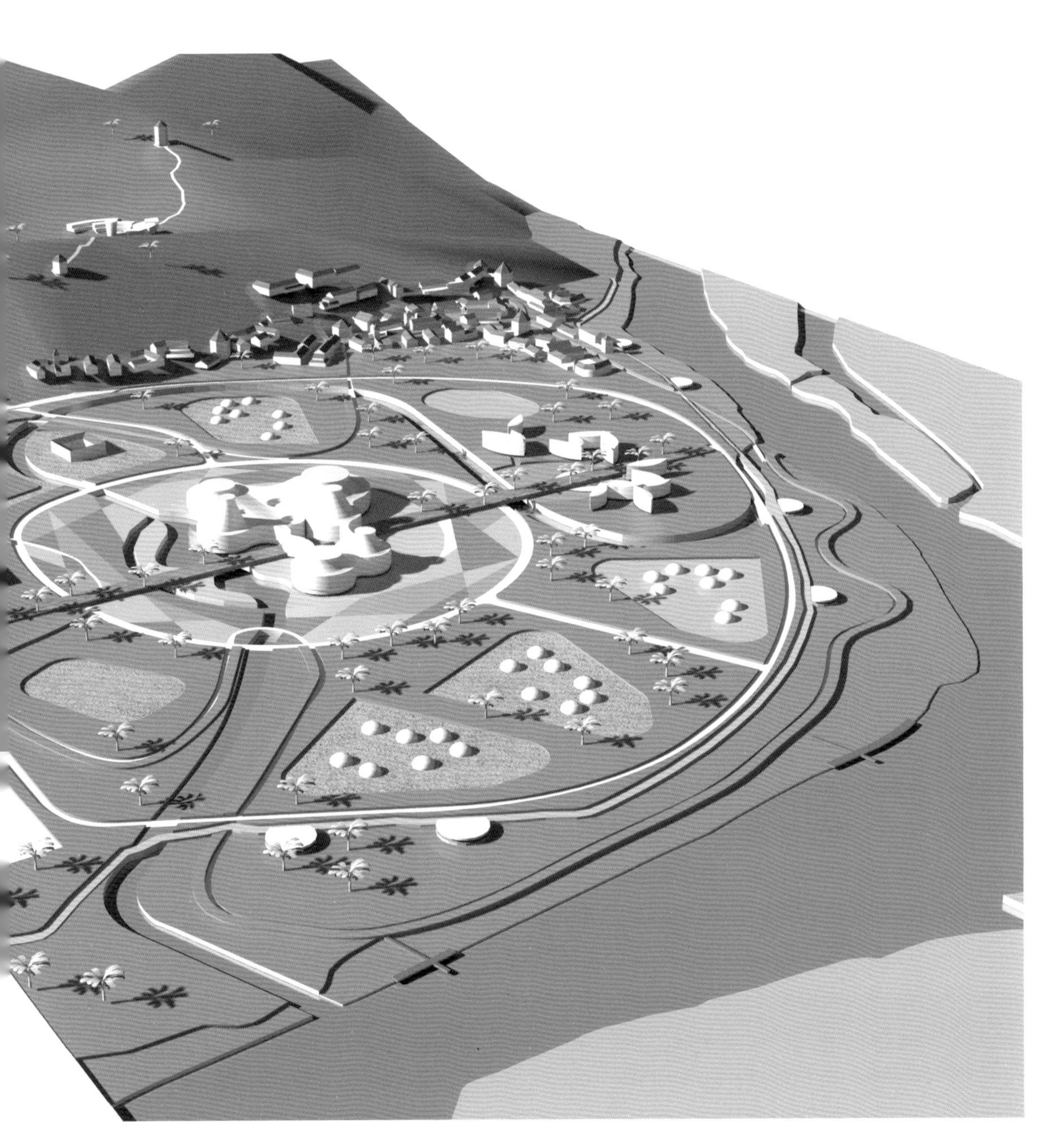

娱乐、管理等6栋建筑组成，建筑造型以蒙古包为原型，融入了敖包、木楞房等其他草原建筑元素。6栋建筑之间用大跨度幕墙结构围合出大型共享空间——亚自然空间，以抵御当地的严寒气候，为冬季游人提供半户外休闲场所。综合体外墙包以格栅表皮，其图案设计来自蒙古包的哈那（围栏）。

Hulun Buir Cultural Park of Nationalities is located in the north of Hailar City, covering a total surface area of one hundred hectares. The park is divided into fifteen sections; exhibitions are mainly composed of minority native experiences panels- such as Mongolian, Ewenki, Olunchun, Daganer, and Russia. Leisure sections offer experiences including skiing, archery, and horse racing. The centre of the park lays a multifunctional complex.

The overall layout absorbs that of a river

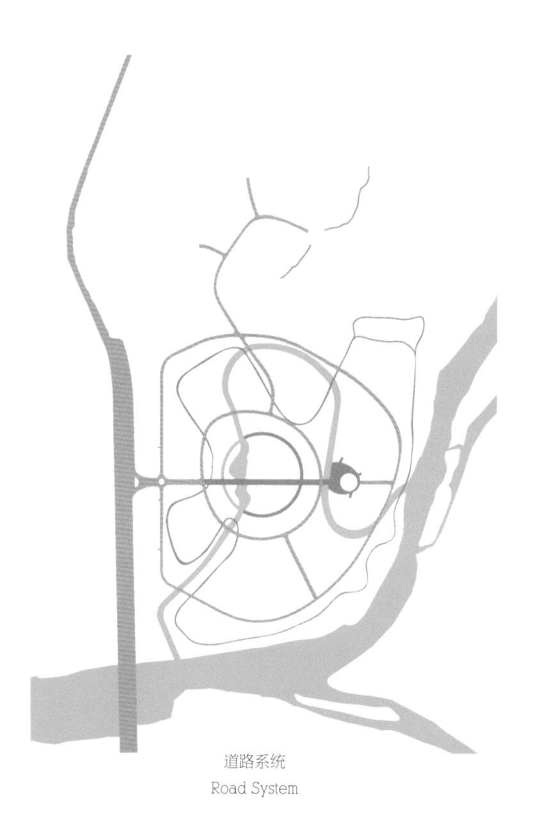

道路系统
Road System

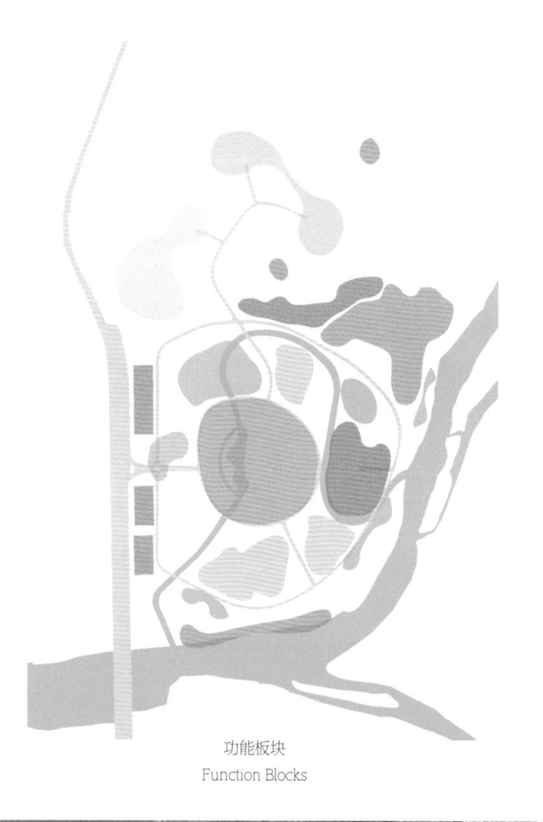

功能板块
Function Blocks

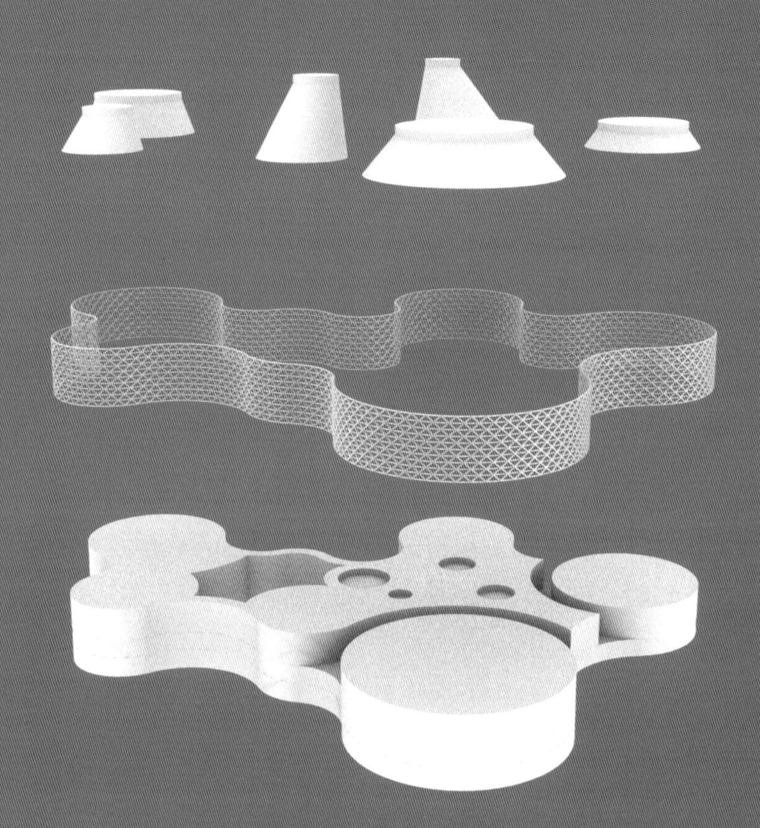

in a prairie, minimizing artificial marks with natural flows to match the original ecological landscape. The multifunctional complex consists of six buildings, each provide facilities for performing art, dining, conference, accommodation, entertainment and management. The architectural designs are based on Mongolian yurts, fused with other elements of prairie such as mounds and batten seams. The six build-

ings utilize a long-span curtain wall structure to create a large shared space, a Sub-Natural Enclosure, to resist against local chilly climate and provide semi-outdoor leisure sites for winter tourists. The exterior wall of the multi-functional complex is wrapped by grating skin; its design pattern is originated from the *hanna*-fence of Mongolian yurts.

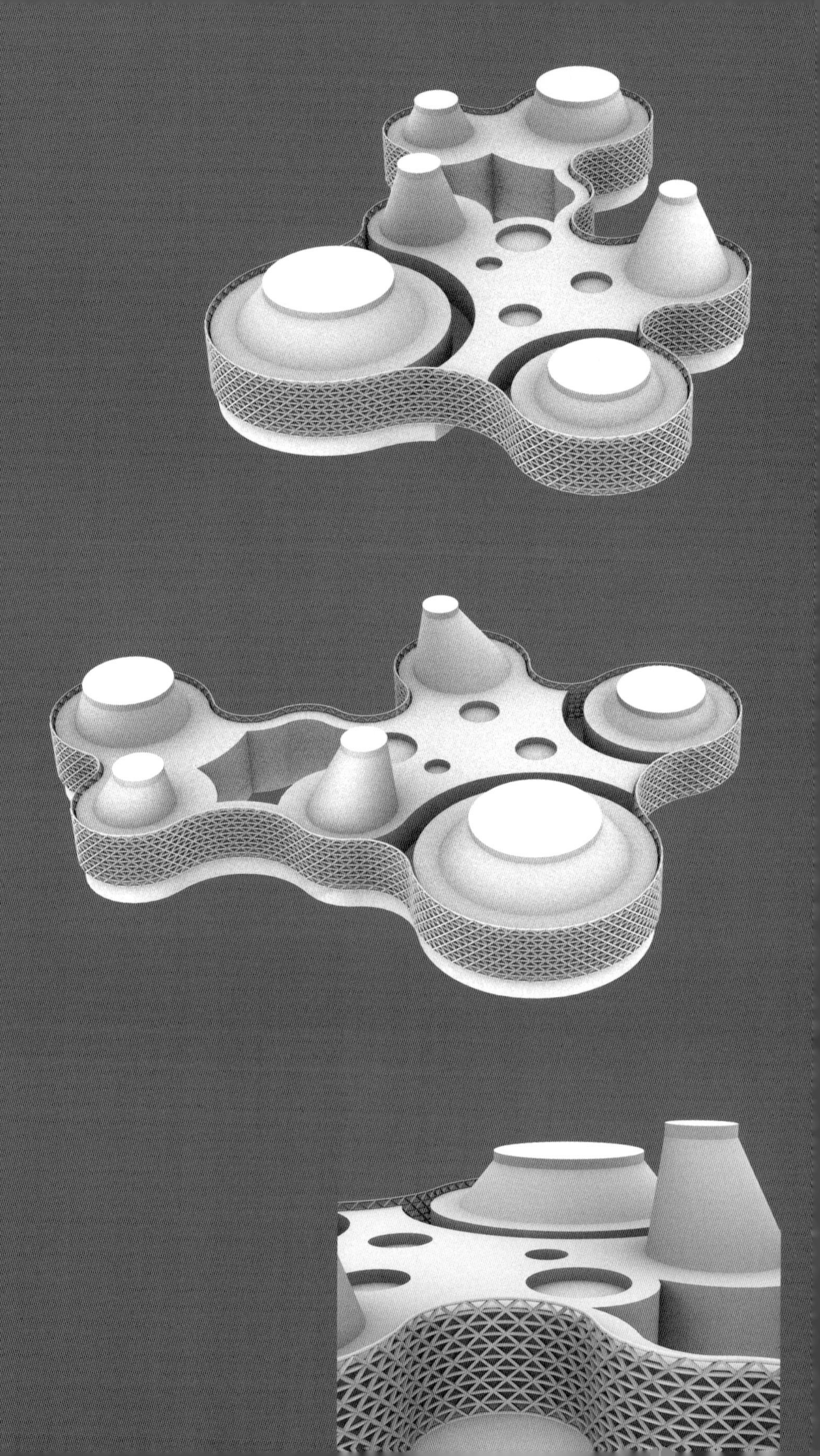

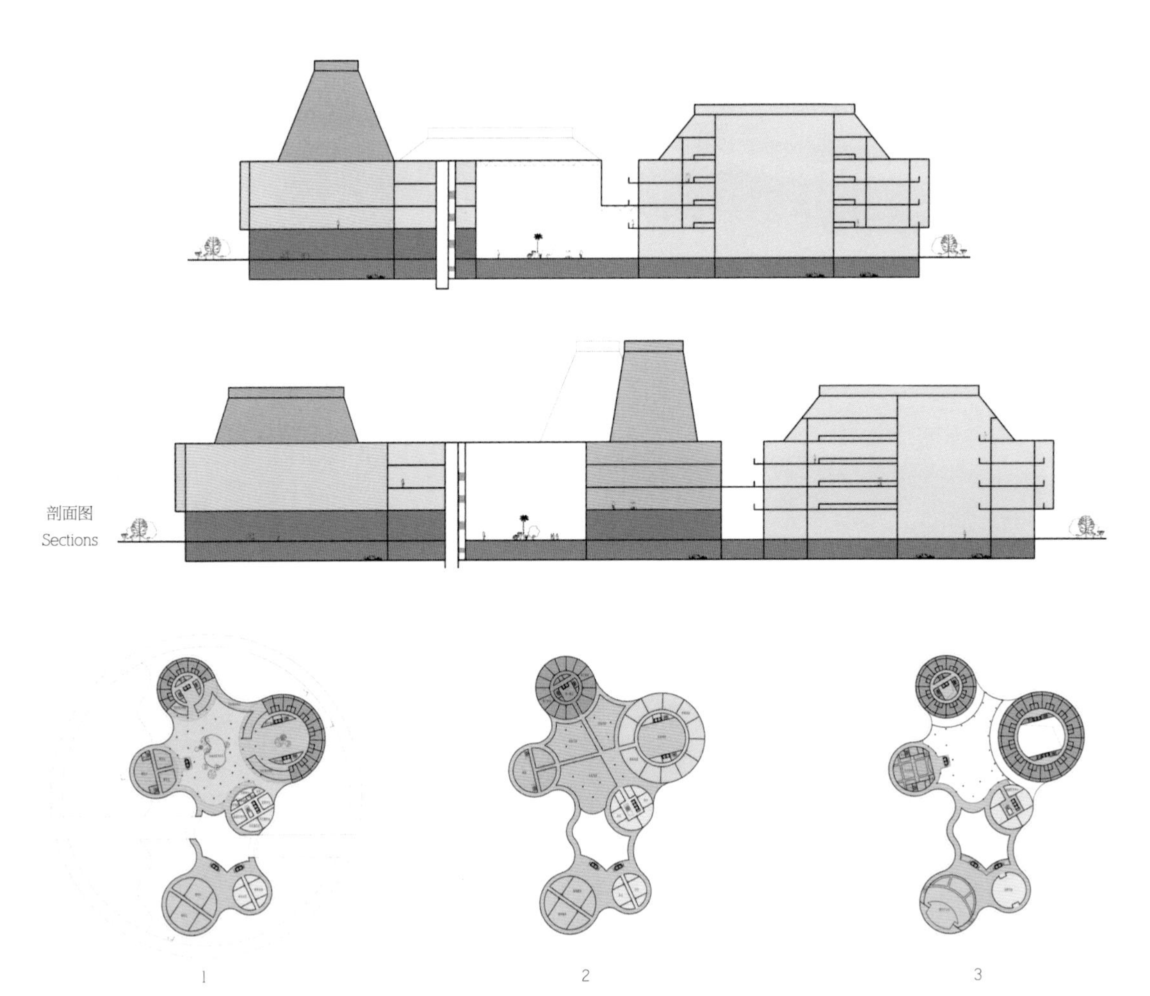

剖面图
Sections

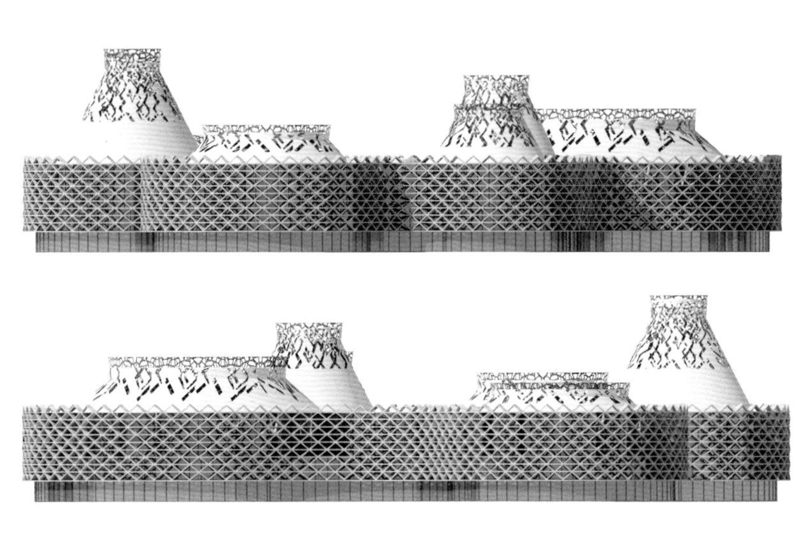

立面图
Facades

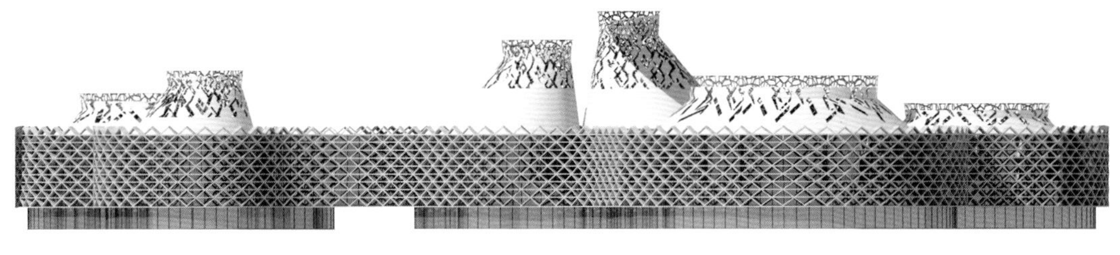

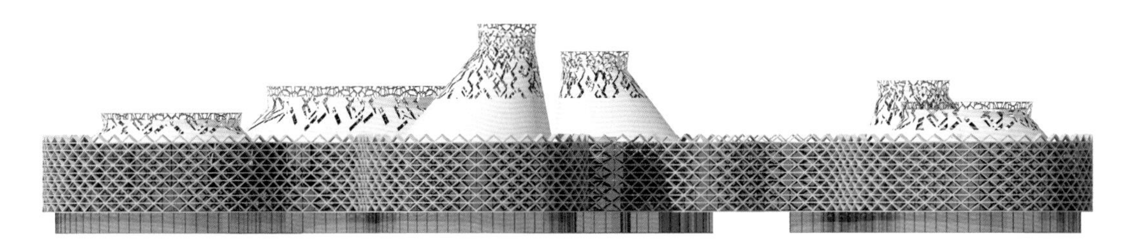

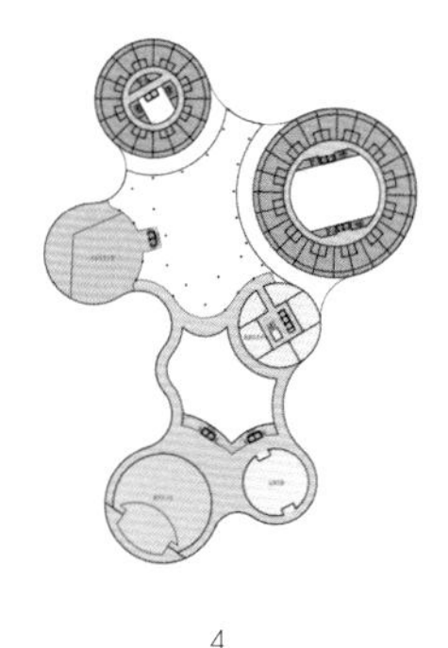

4

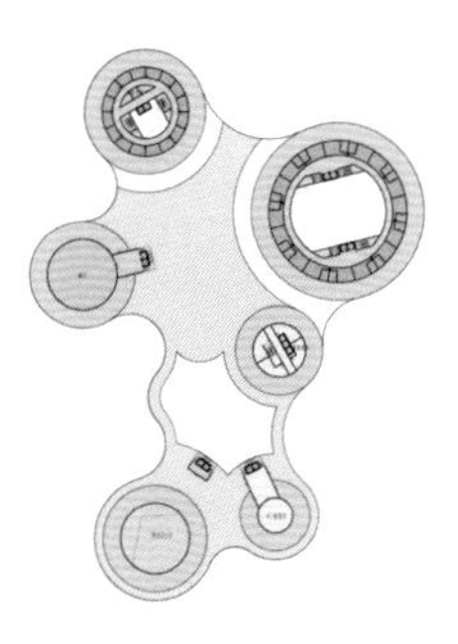

5

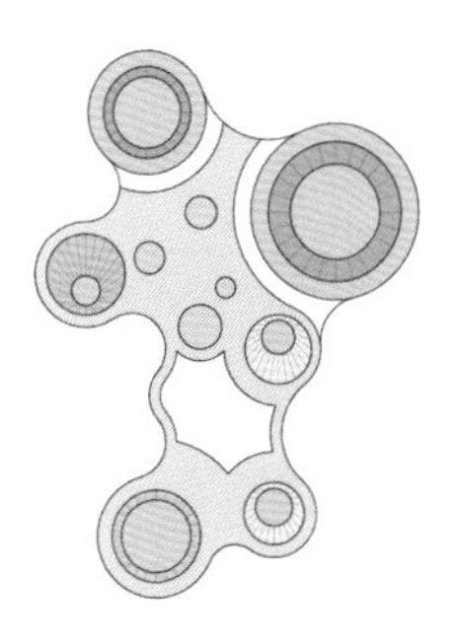

6

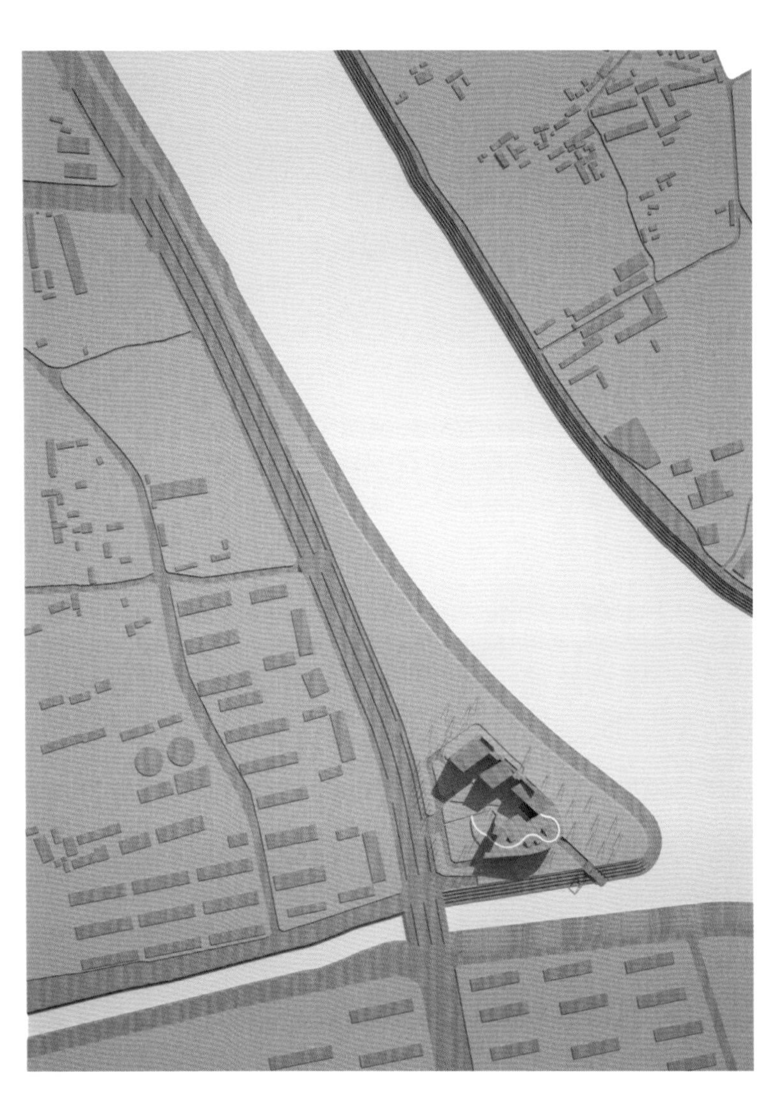

门头沟区规划展览馆

Planning Exhibition Hall of Mentougou

主题：一个触媒

Theme: A Catalyst

该项目位于永定河岸边。永定河是北京的母亲河，历史上由于洪水频发，城市建成区远离了河道，滨水地带成了被遗弃的角落。本项目处于一个特殊的地段，三面临水，具有开阔的视野。建筑师的设计理念是将城市与河流联系起来，让市民可以接近城市的母亲河。

地段内步行道将建筑打穿，将人们引到河边；建筑斜坡屋顶设计成开放的公共平台，可以远眺河面。展览馆的造型来源于建筑与周边环境的互动。

它将成为一个触媒，激发人们回忆起北京的母亲河。

The Planning Exhibition Hall of Mentougou in Beijing rests on the side of Yongding River. Yongding River is the mother river of Beijing, but the city was built away from it for its frequent flooding throughout history; the waterfront became an abandoned corner. The project lies on an exceptional location; it is a peninsula with panoramic view. The architect's design concept aims to link the city and the river, shortening the distance between citizens and the mother river of the city. Pedestrian lanes pass through each building, leading pedestrians to the river. The sloped rooftop is designed as an open platform, giving a glimpse of the river. The exhibition hall's shape resembles the interaction of architecture and

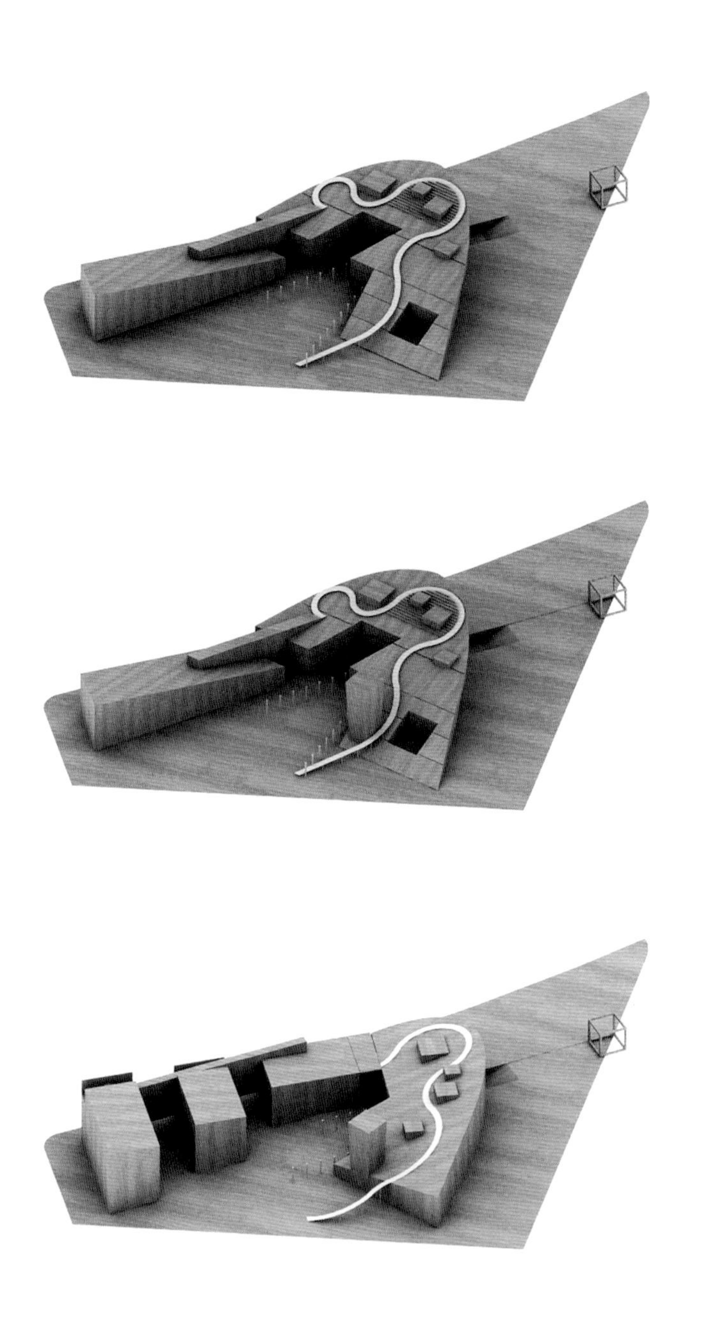
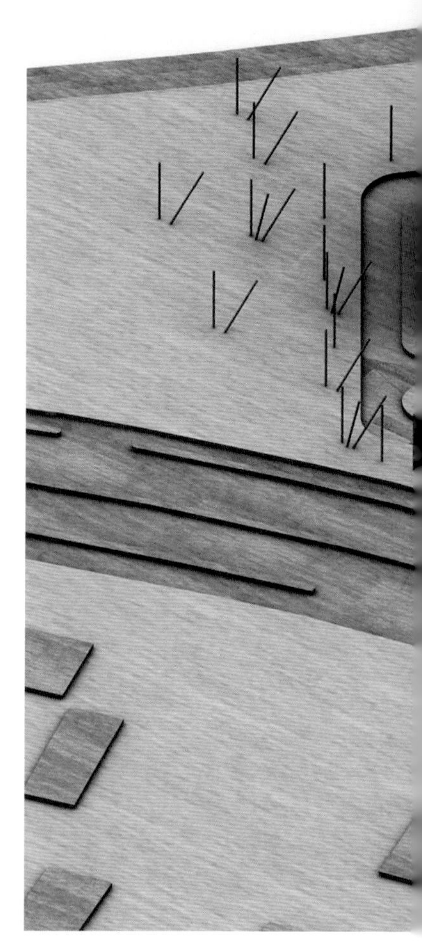

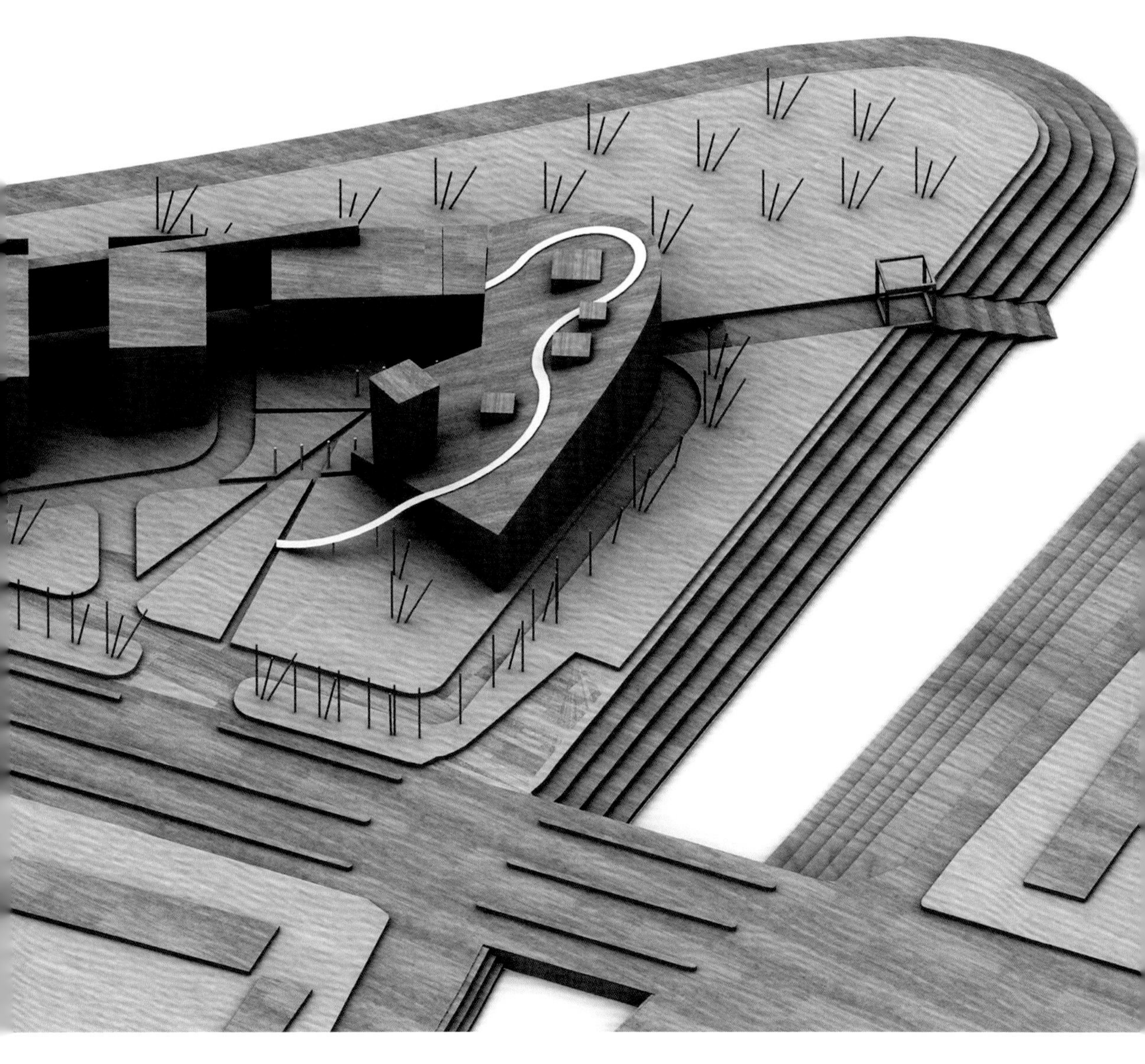

its surroundings. The Planning Exhibition Hall of Mentougou will become a catalyst, to spark the people's memories of Beijing's mother river.

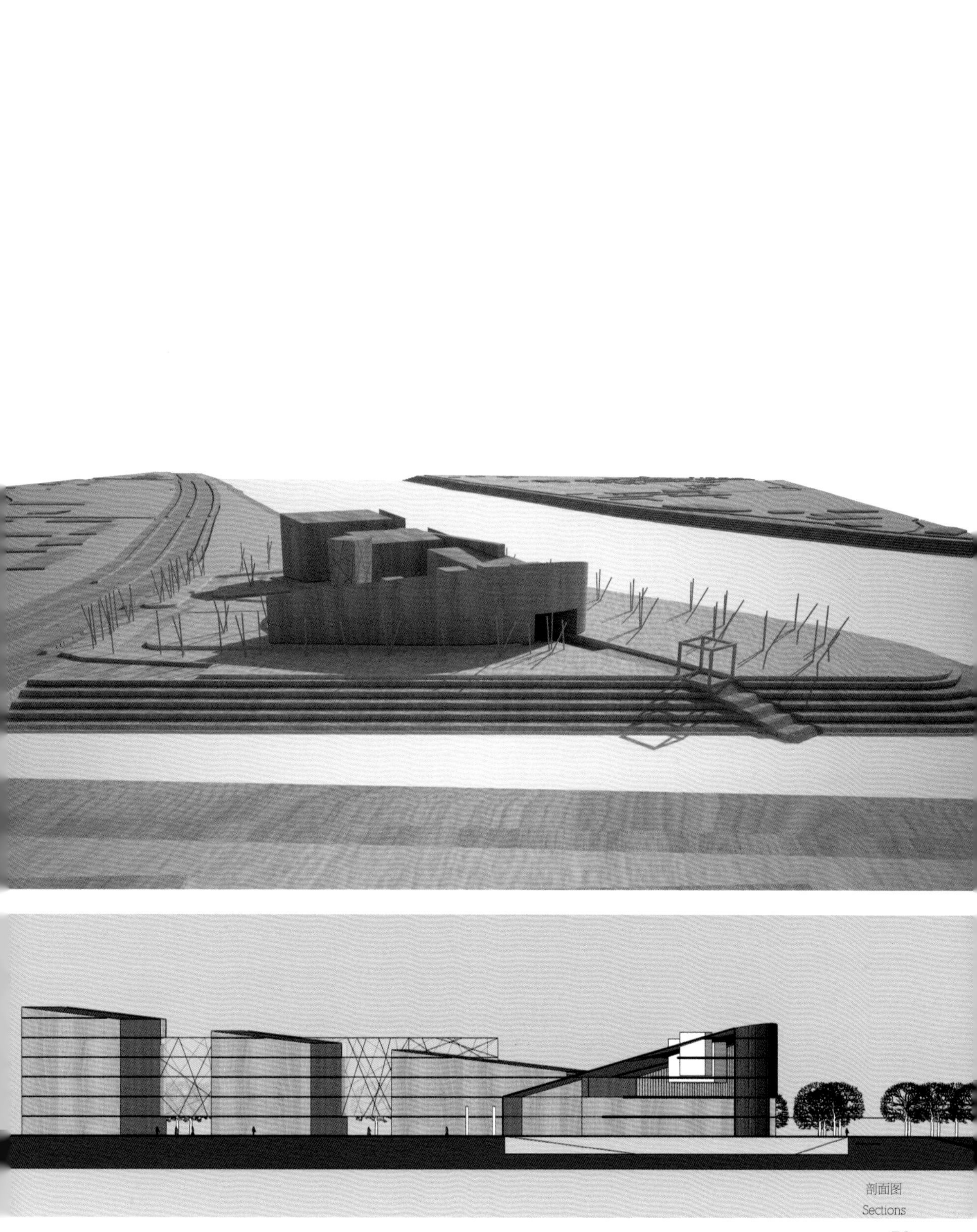

剖面图
Sections

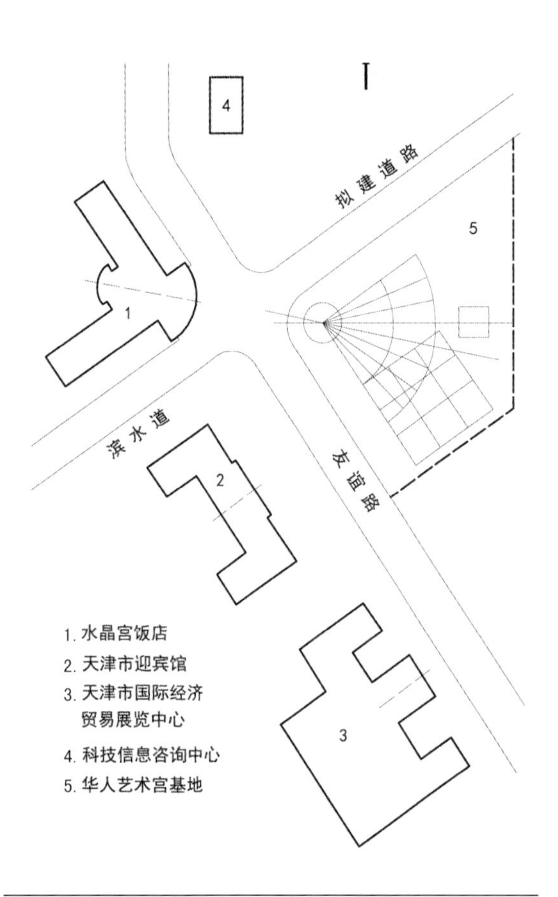

华人艺术宫

CHINESE ART PALACE

札记：我设计华人艺术宫

Note: I Designed Chinese Art Palace

城市中的建筑既是其内部逻辑的自然流露，也是外部条件的客观体现。内部逻辑是使用功能、结构技术、构造作法；外部条件是环境特征、建设法规、历史文化。天津华人艺术宫的创作，除了力争满足以上几点外，还试图通过具有象征意义的建筑空间形态，进一步向人们诉说点什么。

华人艺术宫基地位于天津南郊友谊路与滨水道之交汇处。周围建筑主要有水晶宫饭店、天津国际经济贸易展览中心及天津市迎宾馆。前两者均为近

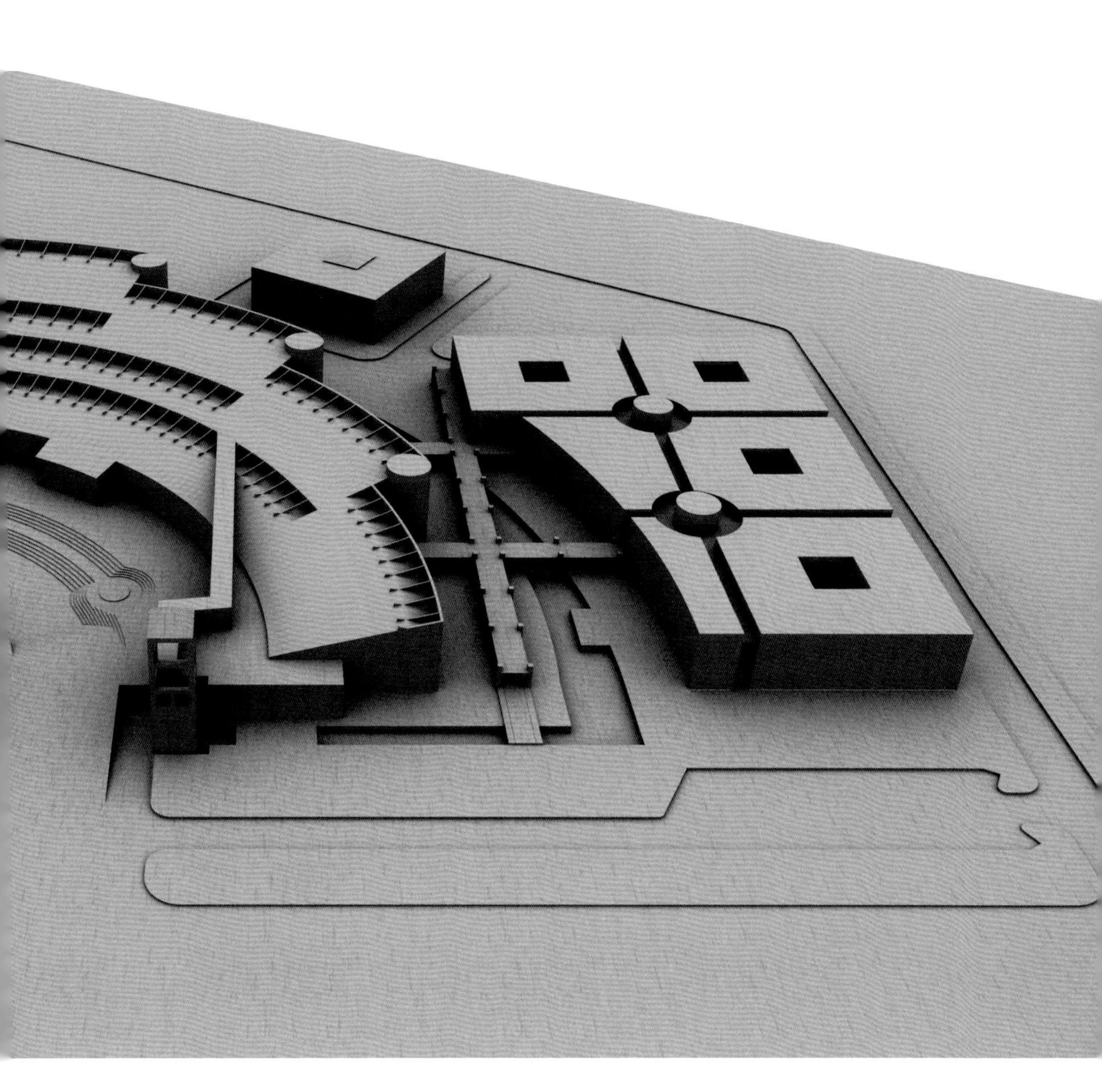

年新作，体量规模庞大，造型完整，采用大片玻璃幕墙，光洁明快，富于现代感。艺术宫用地 6 公顷，建筑总面积 2.5 万平方米，要求分两期实施。一期含现代美术馆、小型实验剧场及演讲厅，二期为博物馆和艺术研究培训中心。另外，要求对室外广场和庭园提出有创见之构想。

从城市设计角度出发，深入挖掘地段特征，力求对地段各项特殊条件以最巧妙地利用是艺术宫创作的出发点。工程基地位于城郊，平整开阔，限制条件较少，粗看起来没有任何约束，似乎可以任意驰骋。但经深入分析轴线关系和地段界线状况，还是发现了“蛛丝马迹”。基地 45 度分角线与正东向线的夹角约 11.25 度，恰为 90 度角的八分角。这些轴线与分角线勾画出艺术宫群体的基本骨架。总体布局恰当地利用和体现了环境所提示的特征并由此展开。

艺术宫总体布局主轴线为何是正东向而非 45 度分角线；扇形主骨架为何分作 8 份而非 9 份，方

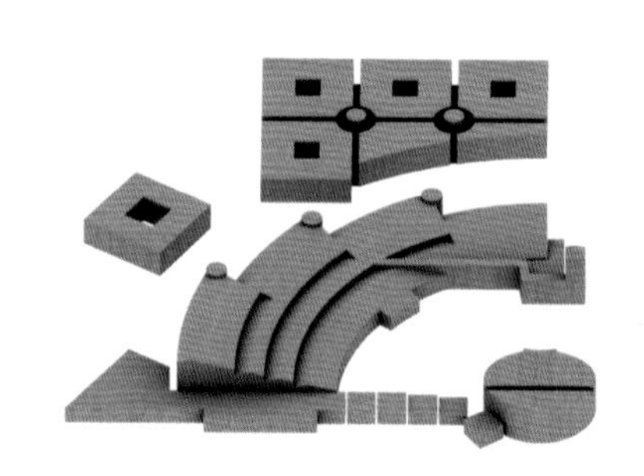

形主骨架为何分为 9 份而非 16 份；这些则是因袭了中国传统的“文王八卦”之说。但这里不是原封不动的照搬，而是将“文王八卦”的两种基本图案，在保持基本特征（正东尊显，中心太极等）的前提下，加以旋转和分离，从而促成了总体布局的确立，并赋予其传统礼仪之色彩。正东为龙之所，至尊显贵，定为轴线并置艺术研究培训中心于其上，暗喻着中华文化兴旺发达。中心太极的一阳一阴则是中国人对全部生活的体验和浓缩，生活本身充满了一幕幕的悲剧喜剧，这里摆上实验剧场也就再自然不过了。“文王八卦”帮助我在轴线多变、地界不规则、建筑群体纷杂的环境中寻得了一种秩序。这种秩序使我放弃了完成一个非南北向四合院的企图；这种秩序帮助我确定了艺术宫在动态中求得平衡的基本设计风格。

另外，总体布局中还考虑了功能分区和建设分期。一期与二期既独立又联系。一期功能和造型相对完整，自身围出广场，在没有二期时也能独立存

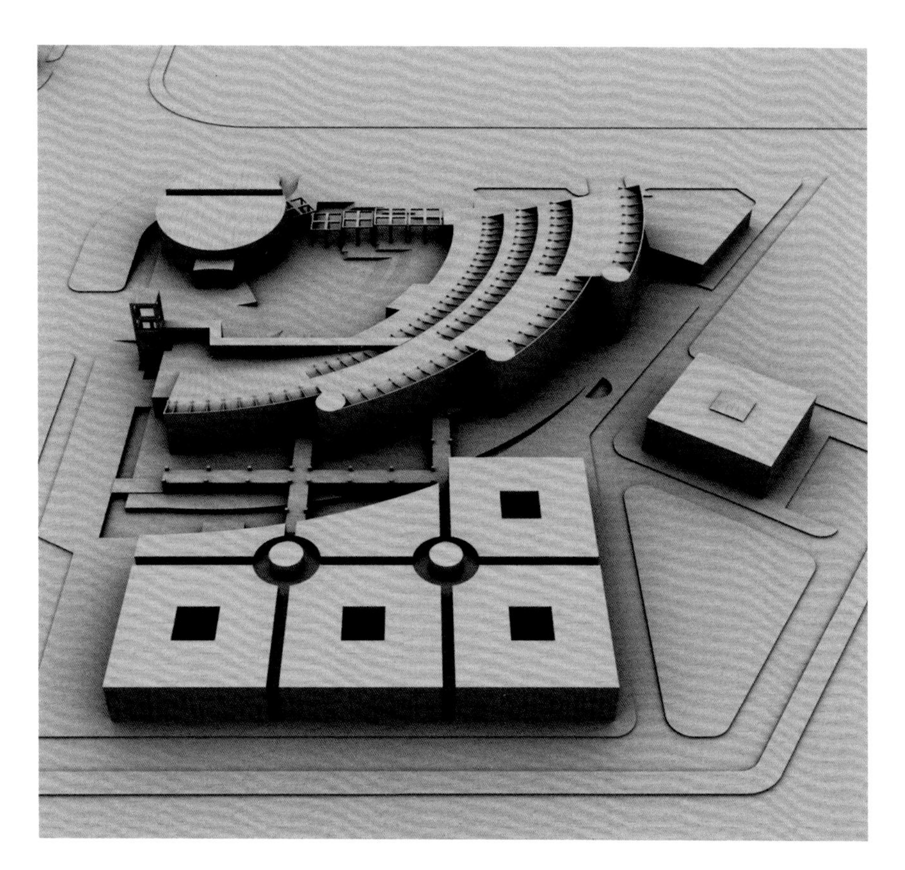

在，给观者以完整的印象。二期建成后，形象上与一期对比统一，并围合出庭园。再者，鉴于当今世界博览建筑之发展趋势，为远期发展留有完整用地。在注重建筑造型完整统一，满足城市空间大尺度要求的同时，利用建筑自身进一步分割室外空间，为市民提供尺度相宜、空间丰富、亲切明快、具有滞留性的室外公共场所。

华人艺术宫作为展示华夏传统文化与现代风貌之建筑，其自身造型也是至关重要的，应力求新颖独到，富于强烈的个性；应具有一定的象征意义，给人以完美的印象和丰富的联想，成为城市风貌中具有标志性的建筑物。

现代美术馆造型宛若一把中国传统折扇。之所以择取扇形，乃因其形象潇洒俊逸韵致风雅。满招损而谦受益，半开半掩之间自然地结合了地形，也含蓄地体现了中华艺术涵虚澹泊之神韵。扇骨焦点之处的实验剧场与之遥相呼应，成就出的一片广场仿佛一露天展台，是室内展示空间（美术馆、剧场）

的延续。在这里，每个人既是观者又是饰者。高耸的钟塔在竖向上丰富了广场空间，并成为一种标志。

博物馆之造型则是借用了中国传统“地分九畴”（九宫、九州）之说。五宫为实，四宫为虚。每一宫再被细分为九格。以此作为中国传统宇宙观具体而微的建筑表达。九宫格代表着规矩，以此暗喻传统；扇面代表着奔放，以此暗喻现代，那么两者衔接之处，观者自然能感受到一种强烈的对比与冲突。虚空间的导入缓解了这一紧张。地面层正东方位（震位，龙之方向）起沿扇形周边引一渠清水，水从龙动；中层空间按九宫骨格架起联系一期二期之间的天桥；上层空间的钢架勾勒出扇面筋骨。以此构成的庭园，空间形态万千，聊以赏心悦目。

细部设计上从古老的象形文字中得到一些启示。古汉字“窗”的多种字形其实正是窗户曾有过的图样的反映，也凝聚了自古以来中国人对窗户的理解。艺术宫的窗形设计正是借鉴于此。从有关描述房屋的古汉字中可以看出，人字形屋架是中国古老建筑之天机。扇形屋面一系列裸露椽梁和椽头既是真实结构的延伸，也是对人字屋架的夸张表达。

加深方案设计深度，注重各项技术条件实施的可能性，亦是方案成败之关键。这主要反映在扇面造型的结构及展室采光问题上。

扇面造型看似复杂，其结构实际很简单。一系列常见的 10 米跨度三角屋架（类似纺织厂房结构）设置在不同高度的柱网上。当其一侧斜面连成一片时，浪漫的大扇面造型即大功告成。在另一侧斜面

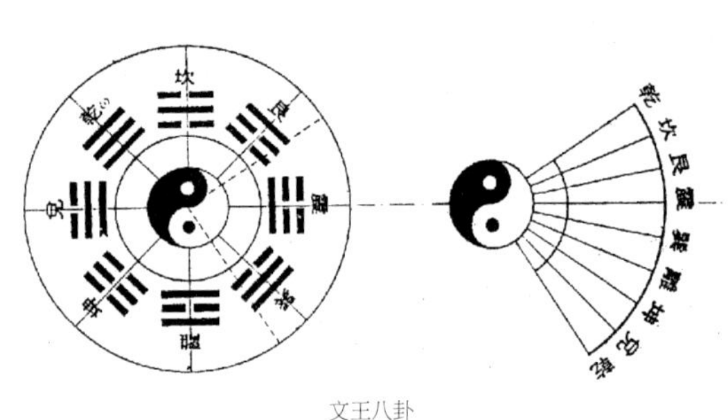

文王八卦
Eight diagrams of King Wen

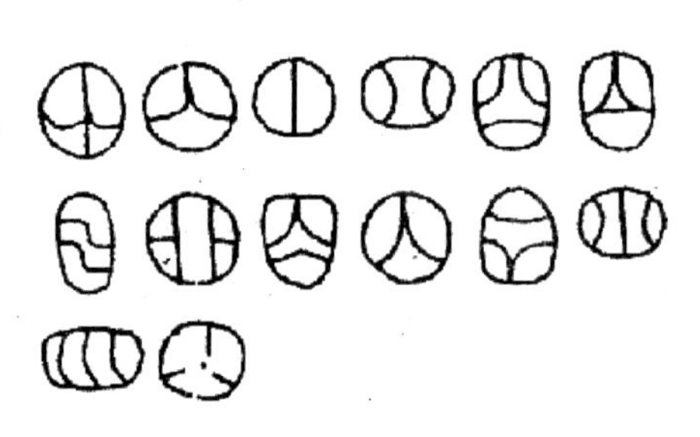

古汉字“窗”
"Window" in classical Chinese

有关建筑的古汉字
Classical Chinese words related to architecture

上设置天窗并架设裸露椽梁于其上。这样不但满足了美术馆采天光的需要，且避免了扇面过于庞大沉闷，透得几分玲珑与轻巧。博物馆则利用天井及侧窗采光。

建筑创作从构思到完成是一个极其艰难的过程。其中的匠心有时就似一个"黑匣"，是很难把它解开说得一清二楚的。对于建筑师来说，最终呈现给广大观者的还应是他的设计作品，并由观者去评说。

现代主义依然是建筑创作的基础。简单模仿传统形式已不合时宜，而应汲取中国传统文化精华融于作品之中。建筑的内部逻辑与外部条件提示了若干可能的基本形式，然而建筑整体形态的产生却是一种超越。

华人艺术宫的创作既是一个逻辑理性的过程，又是一个幻想感觉的过程。它不是老古董出土，不是土特产出洋，也不是航天飞机上天，它应是实实在在可以在当今中国实现的设想。当老百姓惊叹半开半掩的大扇面，甚而称其为"大扇子艺术宫"时；当文化人欣赏"如翚斯飞"的屋面曲线时；当老学究看穿了"文王八卦"构成的礼仪秩序时，华人艺术宫就真实而亲切地呈现出来。

然而扇面也好，九宫也罢，其功能和结构均是理性的。这些空间形态意象是依设计的逻辑产生和存在的。当这些意象为人们所拥有和享受之时，建筑才超越了简单的构筑物而成为一种文化财富。

原文刊于《新建筑》1992 年第二期

Architecture in cities is both the natural revelation of its internal logic and the objective manifestation of its external condition. The Internal logic refers to functional requirements, structural technologies and construction practice; the external condition refers to environmental features, construction codes and historical culture. The construction of the Chinese Art Palace in Tianjin, while meeting all the above-mentioned conditions, also tries to tell people something through its symbolic architecture spatial forms.

The Chinese Art Palace is located at the intersection of Youyi Road and Binshui Avenue in the south of Tianjin. Major surrounding buildings include the Chrystal Palace Hotel, the Tianjin World Economy Trade & Exhibition Centre and the Tianjin Guesthouse. The former two buildings are constructed in recent years with large scale, complete form and huge glass curtain wall, which is bright and modern. The Art Palace covers 6 hectares, with a total floor area of 25,000 square meters, which is constructed in two stages. The first stage includes the Modern Art Gallery, a Small Experimental Theatre and a Lecture Hall, while the second stage includes the Museum and

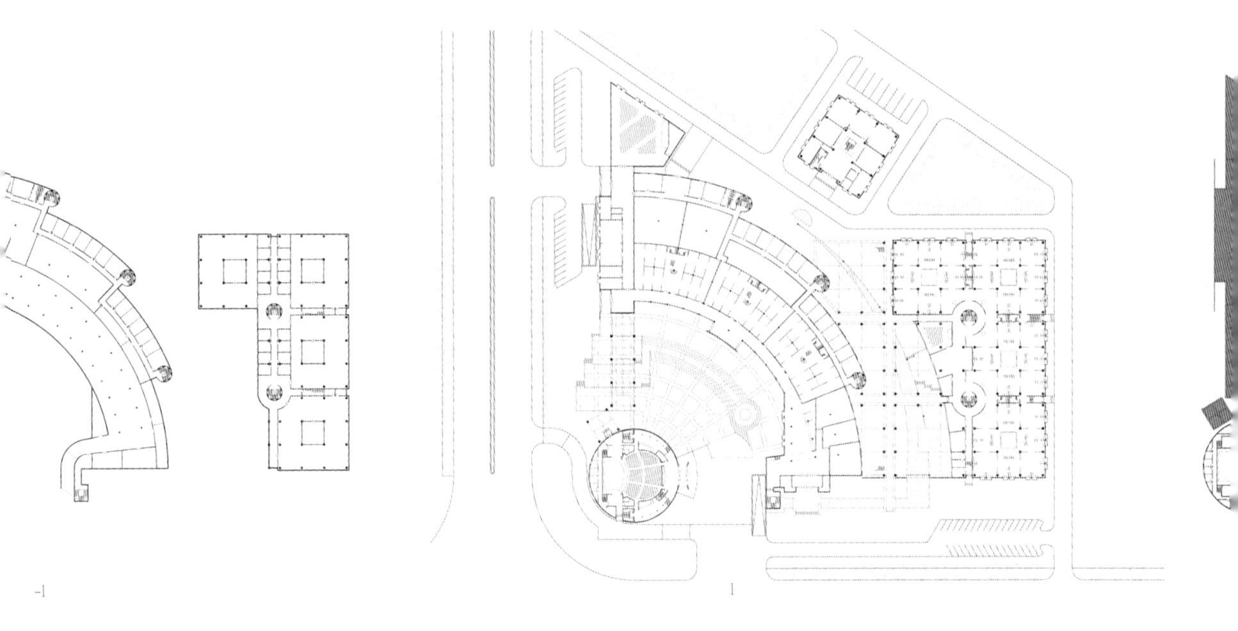

Art Research and Training Centre. In addition, the Palace has some creative thoughts about the outdoor plaza and courtyard.

The starting point of Art Palace's design was to dig deep into local site characteristics and use these specific site conditions as a base for the urban design. At first, because of the project's location in a flat suburban atmosphere, there seem to be no local site characteristics. But after an analysis of the site's axis boundary conditions, we found that the angle between the 45-degree site bisector and the due east line is about 11.25 degrees, which is exactly one eighth of a 90-degree angle. Those axes and angle bisectors then started to delineate the skeleton of the Palace design. The overall layout then properly utilizes and demonstrates the environmental characters in the final building lay out.

But then, why does the principal axis of the overall layout is the due east line instead of the 45-degree bisector? Or why is the fan-shaped main skeleton divided into eight instead of nine parts? And why has the square-shaped main skeleton to be divided into nine parts instead of sixteen parts? That is, because in addition to the site analysis, these conditions follow the rules of China's traditional saying of the "Eight Diagrams of King Wen". Based upon these principles, while preserving the Eight Diagrams' basic characteristics (due east means noble, centre means *tai chi*, etc.), the design rotates and separates the two basic patterns of the "Eight Diagrams of King Wen" to form the overall layout and thus embed a sense of traditional rituals. Due east is the place where

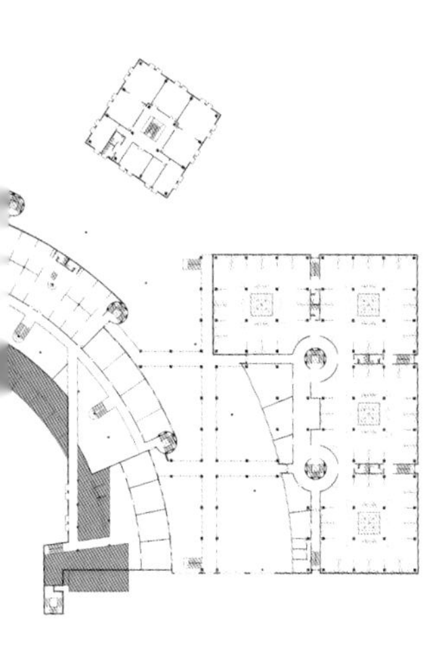

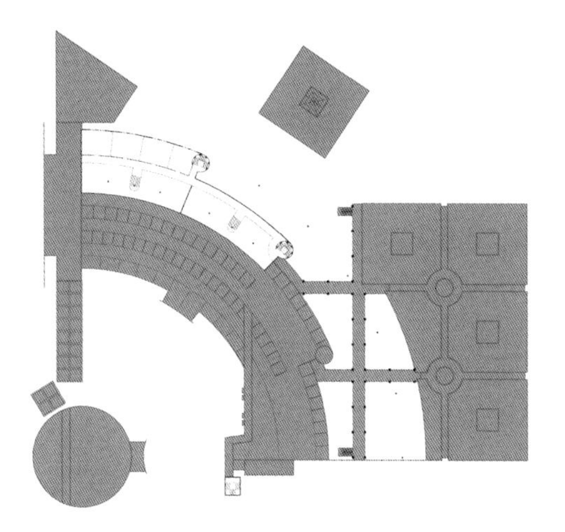

3

4

the dragon stays, which is noble. Then for us, to set the due east line as the axis and then construct the Art Research and Training Centre on it implies the prosperity of China. Next, the *yin* and *yang* of the centre *tai chi* is the experience and concentration of all the life of Chinese people. Life itself is full of both comedies and tragedies, which is very suitable for the placement of the Experimental Theatre. In an environment of variable axes, irregular boundaries and numerous building groups, the "Eight Diagrams of King Wen" helped me find a kind of order. This order let me give up the attempt to build a non-south-north quadrangle dwelling and helped me determine the Art Palace's basic design style which is to pursue a dynamic state of balance.

The overall layout also considers the functional zoning and the division into two construction phases. The first and second phase are both independent and interconnected. The first phase is relatively complete in its form and function, with its buildings circling a plaza. It can exist independently even without the second stage and thus offers a complete impression to its visitors. After the completion of second stage, it is in contrast and unity with the first stage in image and circles a courtyard. In addition, according to the current development trend of world museum architecture, further land has been conserved for future development. While paying attention to a complete and united building form and satisfying the demand for large space in cities, the building itself is used to further divide outdoor spaces to offer an agreeable and comfortable outdoor public place with proper scale, rich spaces and attractiveness.

As a building to show China's traditional culture and modern development, the form of the Chinese Art Palace is also very important. It should be creative and unique with strong character; it should have a certain symbolic significance and offer people perfect impression and rich imagination, so as to be a landmark building in the city's landscape.

The form of the Modern Art Gallery resembles a traditional Chinese folding fan. The motivation for choosing the folding fan as a starting point, is that its image is natural, unrestrained and elegant. Pride hurts, while modesty benefits. This half-closed form fits the landform naturally and implicitly shows the charm of China's art: simple and modest. The Experimental Theatre, located at the intersection of the fan ribs, echoes the Gallery and forms a plaza that looks like an outdoor exhibition stand, the continuation of indoor exhibition space (the Gallery and the Theatre). Here, everyone is both viewer and player. A high-rising clock tower further enriches the plaza space vertically and becomes a landmark.

The form of the Museum also borrows the traditional Chinese saying of "Land divided into nine parts" (*jiugong* or *jiuzhou*). Among them, five parts are real and the other four parts, virtual. Every part is then further divided into nine zones. This is a miniature expression of Chinese traditional cosmology in architecture. These nine parts and zones represent rules, which mean tradition, and the fan represents boldness, which means modern. Then at the intersection of the two, viewers can naturally feel a strong sense of comparison and collision. The introduction of virtual space eases this tension. In the ground, from the due east (*zhenwei*, dragon's direction), a brook is flowing around the fan just like a dragon; in the middle space, overpasses are built to connect the first and the second phase according to the frame of the nine parts; in the upper part , the steel frame is used to outline the ribs of the fan. The courtyard formed in this way has various spatial forms and is a feast for the eye.

In the detailed design, inspiration was obtained from the old calligraphy. The various character patterns of the ancient character "窗" (window) are actually reflections of the past patterns of window and represent the Chinese understanding of the window since ancient times. The window designs in the Art Palace exactly borrow this idea. It can be seen from the ancient descriptions about houses that the "人"-shaped roof truss is the secret of China's ancient architecture. The series of bare rafters and sallies are both an extension of real structure and an exaggerated expression of "人"-shaped roof truss.

Deepening the design and paying attention to the possibility of implementing various technological conditions was also key in the success of this design. This is mainly reflected in the structure of the fan shape and the lighting conditions of the exhibition hall.

At first the fan shape seems complex but its structure is actually quite simple. A series of common 10 meter-span collar roofs (similar to the structure of a textile factory) are set in the column grid of different heights. When one side of the slope is connected, a romantic fan shape is formed successfully. On the other side of the slope, scuttles and bare rafters are set. This satisfies the demand of the Gallery for natural light and also avoids the fan to be too big and boring, thus creating an exquisite Gallery. The Museum relies on the patio and side windows for lighting.

From design to completion, the process of constructing architecture is extremely difficult. The originality in this process sometimes is just like a "black box" which is hard to explain clearly. For an architect, what he presents finally to viewers is his design work which is left to viewers to judge.

Modernism is still the basis of architecture construction. Simple copying of the preceding traditional form is not appropriate any more. Instead, the essence of traditional Chinese culture should be integrated into contemporary architecture. The internal logic and external conditions of architecture provide various possible basic forms, but the construction of architecture's overall form is a kind of transcendence.

The creation of the Chinese Art Palace is not only a process of logic and reason but also a process of illusion and feeling. It is not an antique for excavation, not a local speciality for export, and not a space shuttle for launch. Instead, it is a true vision that can be realized in modern China. When the public is amazed by the half-closed big fan and even calls it "Fan Art Palace", when intellectuals admire the roof curve which looks like "a bird is flying" as described in ancient Chinese poetry and when old pedants see through the ritual order comprised by the "Eight Diagrams of King Wen", the Chinese Art Palace is displayed truly and amiably.

No matter the fan shape or the nine parts, its function and structure are reasonable. These spatial forms and images are born from the logic of design. When those images are owned and enjoyed by people, the architecture can surpass simple structures and become a kind of cultural wealth.

Originally published in *New Architecture*, 1992 (02)

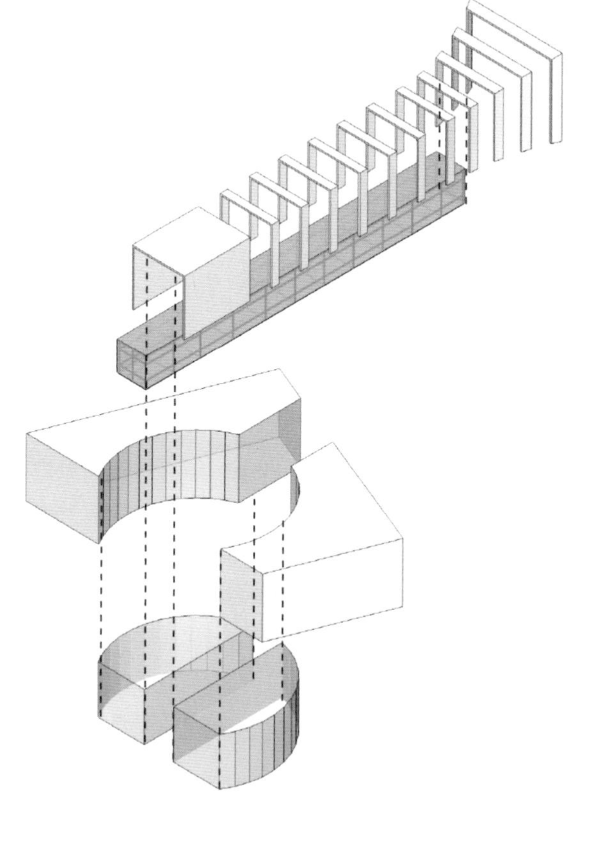

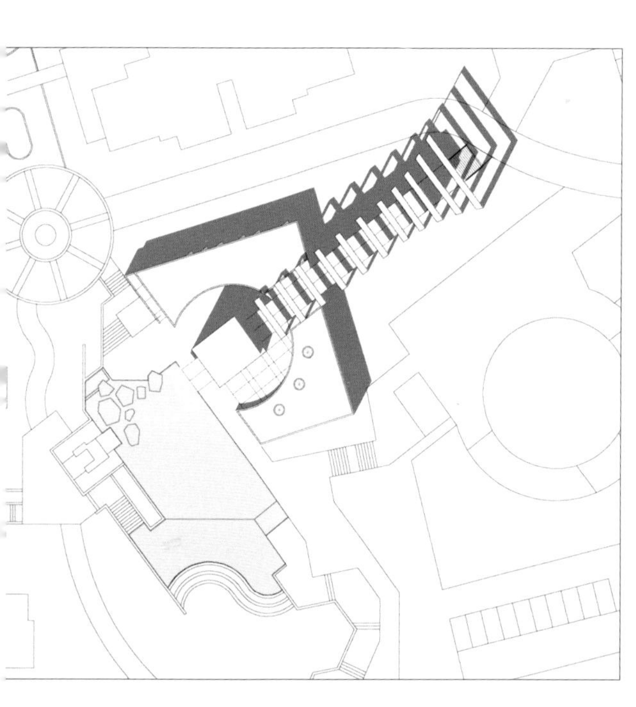

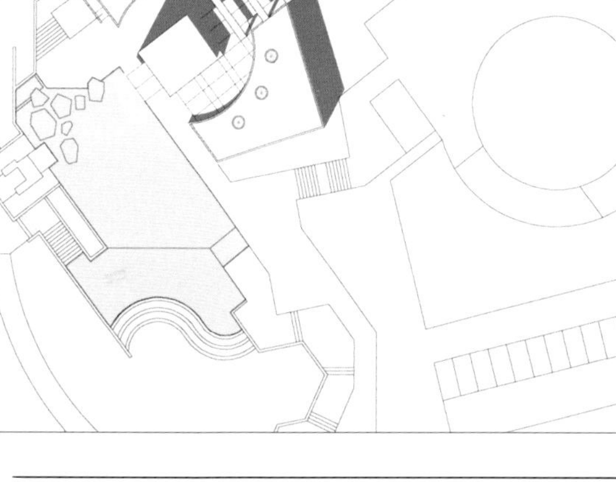

御泉湾会所

Yuquanwan Club

主题：空间过滤器

Theme: The Filter of Space

该项目位于黄山屯溪城区。由于地段有两米多的高差，本建筑面向城市的东北入口设置在二层，而面向社区的西南入口设置在首层。御泉湾会所的建筑形态使用了徽州民居的地域语言，一层层排列的门架借鉴了徽州村落的牌坊群。但对于传统元素，本方案并不是原样照搬，而是应用现代的设计手法重新组合和演绎。在门架中长度达40多米的玻璃长廊，被悬空架起在草地之上。它是一个空间过滤器，让人们完成从浮躁都市到温馨家园的过渡。

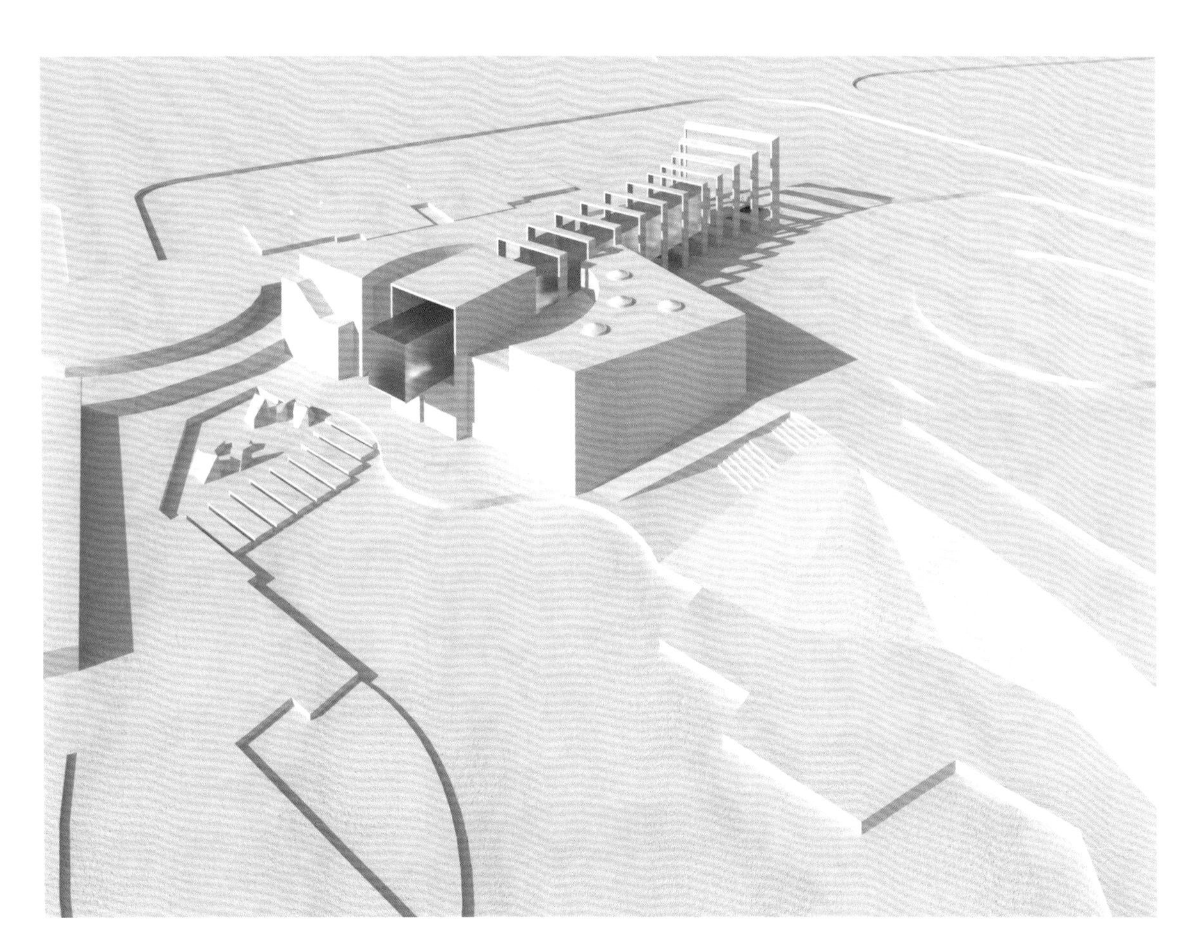

Yuquanwan Club is located in Huangshan's Tunxi district. Since there is a height difference over two meters, the northeast entrance is on the second floor facing the city while the southwest entrance is located on the first floor facing the community. Yuquanwan Club's architectural shape adopts to Huizhou residences' local attributes: an array of doorframes references the memorial archways of Huizhou villages.

This project not only replicates the traditional elements, but also combines and elaborates contemporary design techniques. A glass corridor over forty meters long floats under the doorframe; through the glass flooring shows the green plants underneath. This corridor acts as a filter of space to help people transit from the bustling city to the pleasant home.

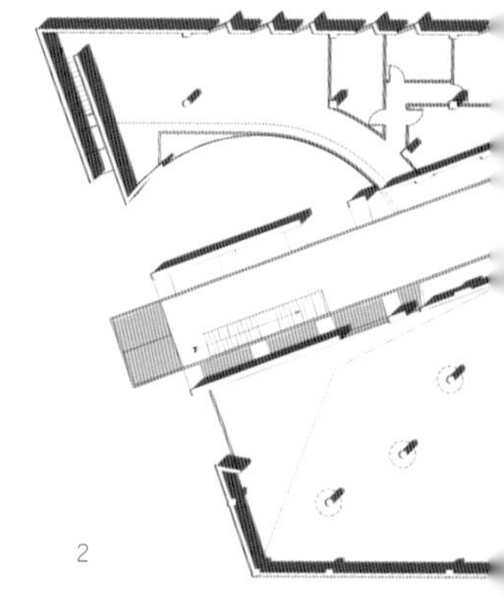

2

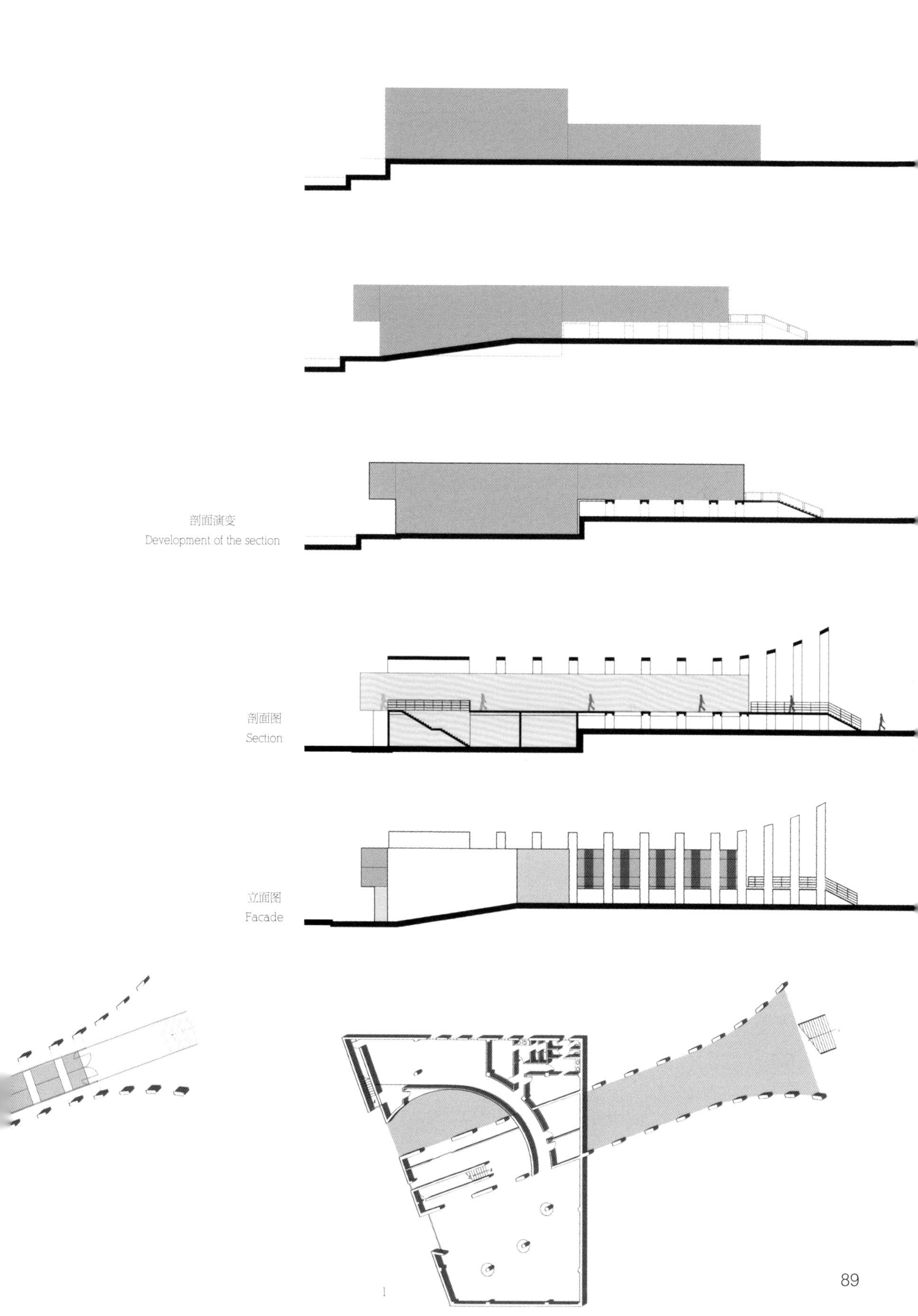

剖面演变
Development of the section

剖面图
Section

立面图
Facade

本土地景　LOCAL LANDMARK

文脉更新　CONTEXT RECONSTRUCTED

本土建构　LOCAL TECTONICS

城市空间　URBAN SPACE

本土再造　LOCAL UPCYCLING

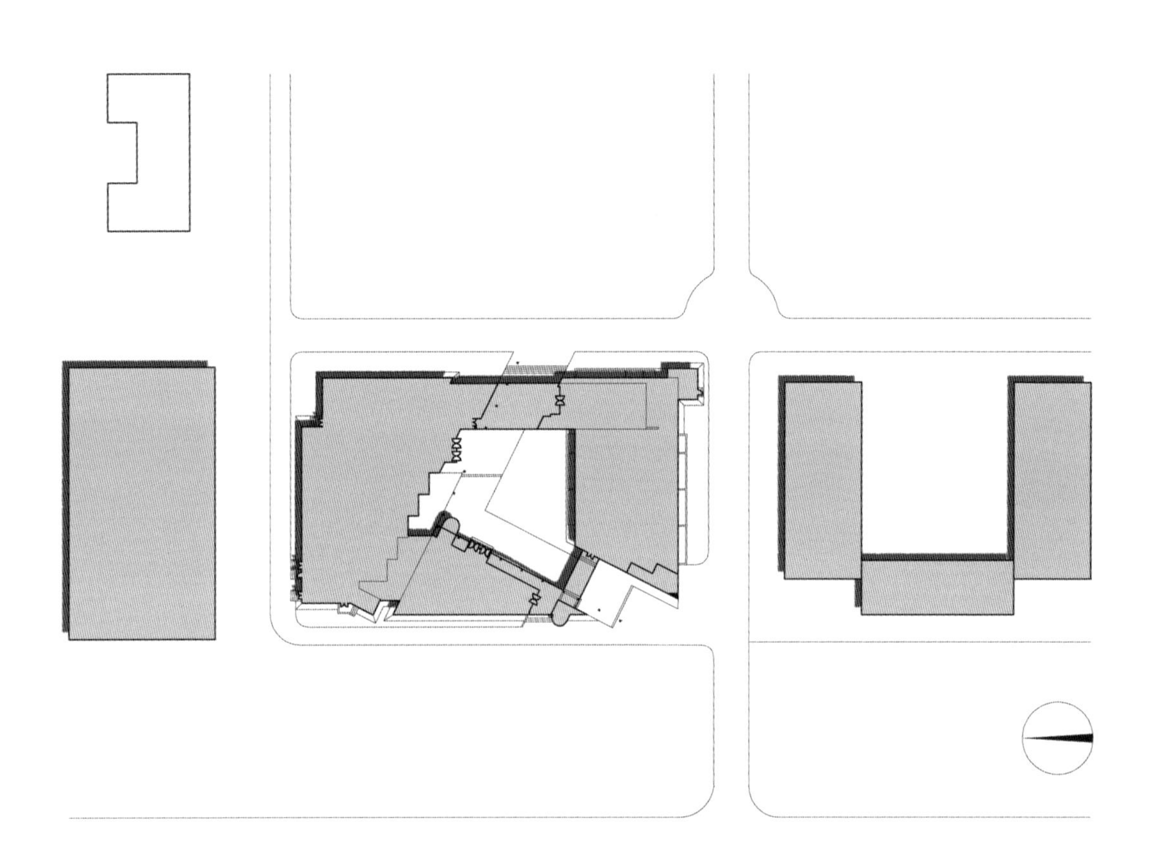

北京电影学院逸夫影视艺术中心

Beijing Film Academy's Yifu Film & TV Art Center

札记：黑白之间

Note: Between Black and White

承担北京电影学院逸夫影视艺术中心的建筑设计，多少带点挑战的心理。电影学院院长带领一干艺术家来清华谈他们的设想，多少也造成点压力。好在艺术是相通的，接下来的设计过程中，作为甲方的北京电影学院在有限的条件下为建筑师提供了广阔的创作空间，给予了充分的理解和信任，也使建筑师有幸将一些设计理念付诸实现。

北京电影学院逸夫影视艺术中心紧临学院路，与重建的“燕京八景”之一——“蓟门烟树”隔路

相望。该建筑由教学楼、公寓楼及餐厅组成。除了后退红线(set-back requirement)、建筑限高(height limitation)等通常设计难点外，该项目的设计难点还有避免西晒，以及使建筑从西侧学院路和东侧校园内部都能方便进入。

当然这并不是问题的全部。建筑的内部功能和外部条件仅仅提示出建筑的某些基本可能，而建筑最终形式的产生，则需要建筑师的创造性劳动。其背后反映出的是建筑师的学识、品味及追求。围绕

北京电影学院逸夫影视艺术中心的设计理念，这里从四个方面加以论述。

·“有序”与“礼”

在建筑设计中，我通常追求一种“有序”的感觉。这种“有序”不一定是唯一的，但是有理性基础的。这种“有序”来自某种暗示，可以是周边环境，可以是建筑自身，可以是某种理念，也可以来自某种技术，但更多的是综合作用的结果。综合则需要建筑师的判断与把握。这种“有序”不是平白直叙的，但又是易于理解与掌握的。这种“有序”在某些方面可理解为中国传统文化中的“礼”。

《礼记·经解》中曰：“礼之于正国也，犹衡之于轻重也，绳墨之于曲直也，规矩之于方圆也。”“礼”之于中国不仅意味着天人关系和伦理道德，也意味着生活行为及经国方略，当然包括营建规范。尽管我们今天的“礼”与传统的“礼”已有所不同，但讲“礼”、重“礼”仍应是我们这个“礼仪之邦”的崇尚。在建筑界，动辄“地标式建

筑”、“二十年不落后”的提法，显然与这种“礼”是不相宜的。

· “黑”与“白”

读过美国学者 N. Eshan《空白的质量》（《建筑学报》1998 年 1 期）一文很有感触。西方人对“白”的理解，尤其是建筑上“空”的理解并不逊色于东方，而传统上这应该是中国人的长项。中国传统美学中，无论是绘画的“象外之象”，诗文的“无言之境”，还是音乐的“弦外之音”，讲究的都是对“虚”、“空”的利用，也就是我们常说的“计白当黑”。建筑界人士常引用老子《道德经》中的论段，“三十辐共一毂，当其无，有车之用。埏埴以为器，当其无，有器之用。凿户牖以为室，当其无，有室之用。”这里讲的是“有”与“无”的辩证关系。庄子《齐物论》中也有类似论段：“得其环中，以应无穷。”讲的则是“白”的妙用。

对比 1748 年 G·诺利罗马万神庙街区地图和乾隆北京鼓楼街区地图可以看出：无论中西方，院

落、街道、广场都是构成城市“空白”空间的重要类型。不同的是中国的院落空间是积极的，街道空间是消极的；西方院落空间是消极的，街道空间是积极的。传统上中国城市较少有广场，仅有的一些集市广场或宫廷广场也是院落空间的扩大或延伸。如古长安有繁华的东市与西市，各市有市墙围绕，称为阛，市墙有门，称为阓，供买卖人出入。明清北京承天门（天安门）前开有“T”形宫廷广场，广场东西南三面筑有宫墙，正南开有大明门，东侧开有长安左门，西侧开有长安右门。西方城市广场则多是街道空间的延伸及扩大。被称作欧洲“起居室”的威尼斯圣马可广场，一面向大海敞开，其他面通过众多街道与城市相连。此类起居室式的广场在意大利古城中星罗棋布，在市井生活中扮演着重要角色。而西方四边形住宅街坊中心的院落大多仅供采光、晾晒衣物，甚至堆放杂物使用；面向街道的立面装饰考究，面向内院的立面则要简陋得多，不像中国的四合院住宅考究的立面朝向院内。

作此中西比较是要强调对一种中国传统建筑空间的重要类型——“积极的院落空间”的探索。这种空间作为一种类型，对今天的建筑仍有价值。尤其在当今，建筑前的草坪（只能看的）及广场（只能站的）愈发成为城市新宠的时候，这种尝试是不是更有了另一番的意义？“积极的院落空间”，核心是“积极的”，而无论这个内院是否是四合的，或者是否有大屋顶。

· 意境与“蒙太奇”

中国的艺术传统注重于意境的创造。唐代诗人王昌龄在《诗格》中举出诗有三境，“一曰物境”，“二曰情境”、“三曰意境”。宋代诗人苏试在评价王维的诗和画时谓：“味摩诘之诗，诗中有画，观摩诘之画，画中有诗”（《东坡题跋 · 书摩诘〈蓝田烟雨〉》），说的是诗寓情于景和画借景写情。空有技能而无意境，在中国的传统美学中往往被视为雕虫小技而加以摒弃。中国人对知觉的认识本应高于西方，我们的先辈很早就用大理石作挂屏，并在

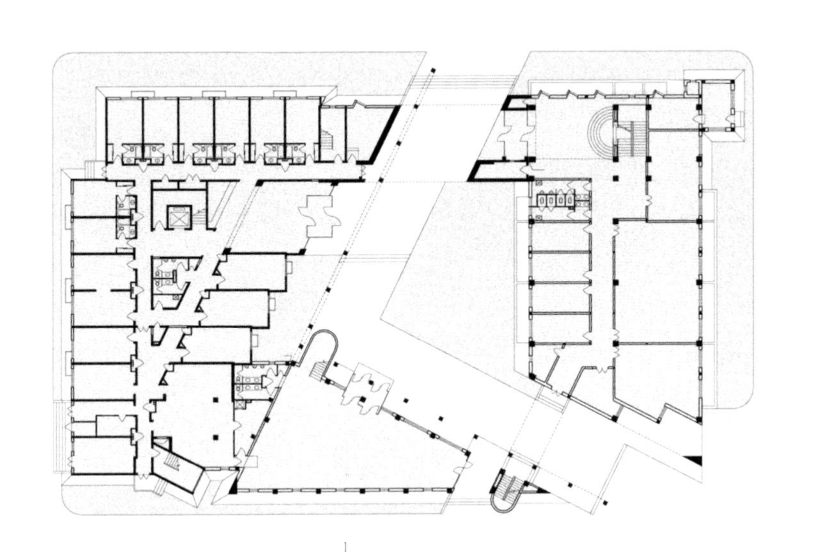

1

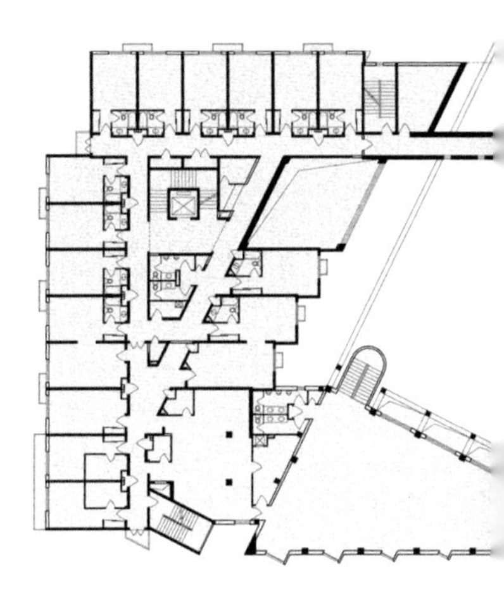

2

石质的纹理中表达与理解涵义。黑格尔的“以形写神”，比中国晋代顾恺之的“迁想妙得”要晚1500年。

意境是中国传统审美中的要素，即使在技艺性很强的建筑、尤其在古典园林中表现得也很突出。具体到意境的创造可以看到如下特点：（1）强化“时间”：在建筑第四维空间（时间）创作上，中国传统建筑有其独特之处。无论是皇帝的紫禁城和天坛，还是江南的私家园林，时间都成为空间意境创造的重要因素；（2）弱化“单体”：在建筑三维空间创作上不追求新奇。既使是太和殿或祈年殿这样的单体，相比西式建筑也要平淡弱化的多；（3）虚实相生：如沈复在《浮生六记》（*Six Chapters of a Floating Life*）中说：“大中见小，小中见大，虚中有实，实中有虚，或藏或露，或浅或深，不仅在周回曲折四字”也；（4）文本相佐：中国人善于把物质性最强的艺术形式——建筑，与精神性最强的艺术形式——诗词联系起来。一方匾额两幅楹联往往能成为建筑意境的点睛之笔。

意境的产生并不是不同意象之间的简单组合，它有些类似电影的“蒙太奇”组合。两个镜头的“蒙太奇”组合，不等于两个镜头之和。它所产生的新的意象境界，远远大于原有意象。即中国美学中常说的“象外之象”。

中国传统建筑中的意境，对我们今天的建筑创作尤其是城市设计仍有很好的借鉴意义。我们在学习西方的时候，不要忘记自己还有这么有价值的传统，将其融入今天的城市是建筑师的职责。不求形体的新奇，但求空间的意境。看到如今越来越混乱的城市风貌以及对所谓“标志性”、“陌生感”之形式的追求，“意境”已离我们越来越远了。

· “非物质化”与情感化

随着信息时代的到来，一种新的理论开始出现，即建筑的“非物质化”（dematerialization）。所谓非物质化是相对于传统建筑材料砖、石、木来说的，即用玻璃、钢及金属板等新材料，表现一种“非物质化”的概念，或“非传统物质”的概念。这也自

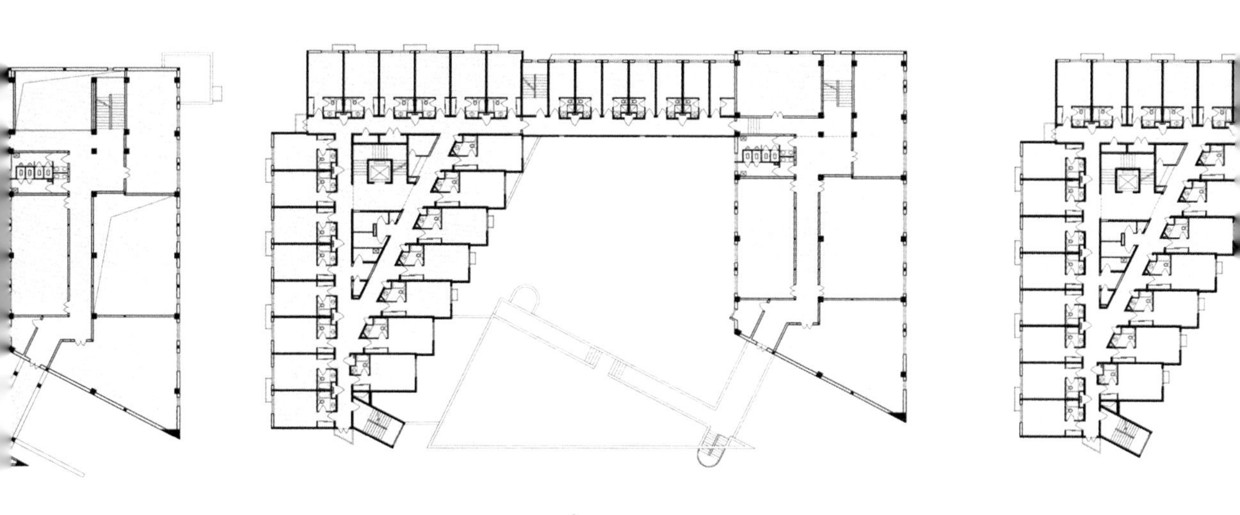

3

然带来有别于传统建筑的审美转变。所以现代社会上有“要想新，就用金（属），要想酷，就用（玻璃或金属）幕”的说法。但无论是“物质化”的砖 + 石 + 木，还是“非物质化”的玻璃 + 金属，都不是建筑的最终表达。形式与材料背后蕴藏着的是人的情感。

印度建筑师柯里亚（Charles Correa）谈自己的创作体会时说，建筑师应该从东方文化中发现神秘而美丽的东西用现代的技术表达出来。庸俗的建筑之所以庸俗，是因为它与人的内心毫不相干。柯里亚是一位善于用“物质化”材料表达情感的大师，而那些善于用“非物质化”材料表达的大师与以往相比也发生了不小的变化。看看意大利建筑师皮亚诺（Renzo Piano）的新卡里多尼亚吉巴欧（Tjibou）文化中心，没有了早期高技派建筑（High-tech building）机器般（machine-like）的冷冰冰，摇身变成蓝天与大海之间的舞者。他说：“建筑真正意义上的广泛性应通过寻根、通过感激历史恩泽、尊重历史文化而获得”。

“物质化”或“非物质化”并不是问题的关键。一位来京的法国建筑师自己跑去电影学院看逸夫影视艺术中心后说，谁说灰砖盖不出好房子，指的就是这层意思。

透过以上四个方面的论述，观者可以更好地体会建筑师在北京电影学院逸夫影视艺术中心设计中的种种匠心和追求。但还是那句老话，见之于形，感之于心，最终说话的还是建筑作品本身。

因为这项工程，我成了北京电影学院的常客。工程还未完全竣工，就开始不断遇到学生来工地实习拍摄。与他们聊天中，学摄影的说，这栋建筑有非常丰富的角度，许多场景可供拍摄选择；学表演的说，建筑的场景营造出个性化的气氛，可烘托他们的气质和表演；一位学声乐的则兴奋地告诉我他要在这里拍 MTV 的计划。学生们的喜爱就是对建筑师的最大奖励。

原文刊于《建筑学报》2000 年第十期

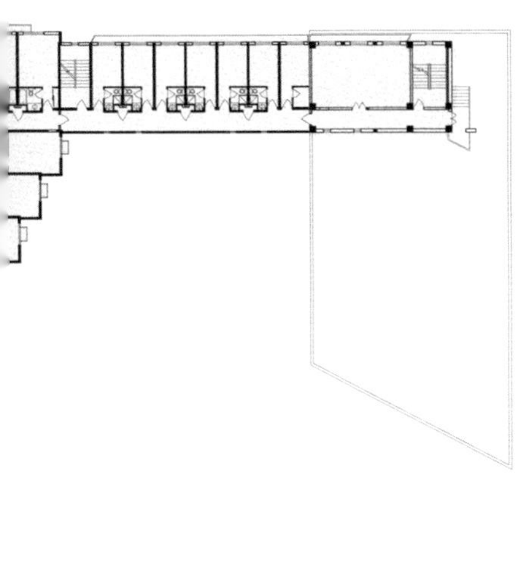

4

The task of designing the Beijing Film Academy's Yifu Film & TV Art Centre was a bit of a challenge for me. There was additional pressure on me from a group of artists (headed by the President of the Academy) who came to Tsinghua University to discuss their plans for the centre. Fortunately, art has always been interconnected. In the subsequent design period, limited by strict conditions, the client nevertheless provided wide room for creative thought and showed sufficient understanding and trust which allowed the architect to realize the design concept.

Bordering XueYuan Road, the Beijing Film Academy's Film & TV Art Centre is situated directly across the road from "Jimen Yanshu"– one of the Eight Views of Yanjing–currently under reconstruction. The Film & TV Art Centre

is comprised of a teaching building, an apartment building, and a cafeteria. Aside from the common practical limitations of architecture, such as set-back requirements and height limitations, additional unique design considerations included avoiding a western exposure and securing access to the complex from both Xueyuan Road in the west and the campus in the east.

Of course, these are only a few factors affecting the ultimate design outcome. Since programmatic requirements and site conditions can only offer a limited set of clues about the final possibilities of a design, the end product calls for creative contemplation on the part of the architect. This is ultimately a reflection of the architect's knowledge, taste, and goals. The following discussion will elaborate upon the design philosophy of the art centre from four aspects.

· *"Order" and "Li"*

Throughout the design process, I often seek a sense of order. This "order" does not have to be exclusive, but should be based upon reason. It stems from hints coming from either the surrounding environment, the architecture itself, or a certain concept or technology, although, more often than not, it is the outcome of a synthesis of all these factors. Such a synthesis requires the architect's proper judgment and choice. Nevertheless, this "order" should not be simplistic or two-dimensional, though it should simultaneously remain easily comprehensible and controllable. In some aspects, such "order" can be understood as "*Li*" in traditional Chinese culture.

The chapter "Different Teachings of the Different Kings" of the *Liji* says: as measure is to weight, a carpenter's ink marker is to lines, a compass and a carpenter's square is to circles and squares, so is *li* to the regulation of a country. "*Li*" in Chinese culture is not only a token of the relations and morals between man and nature, it also stipulates proper behaviour and the governance of a country, including the norms of architectural design. Although today, the concept of "*Li*" is interpreted differently from what it was traditionally, the emphasis of it remains in our highly esteemed "state of ceremonies". Today's frequent claims of landmark designs or boasts of permanently fashionable buildings in the circle of architecture are incompatible with the concept of "*Li*".

1748 年古罗马万神庙街区地图
1748 map of Rome's Pantheon district

· *Black and White*

The American scholar N. Eshan's article "The Importance of Emptiness and Blankness" (published in issue 1,1998 of the *Architectural Journal*) set me thinking. The Western interpretation of blankness, especially on emptiness in the field of architecture, is no less profound than that of the Chinese, even though traditionally the Chinese people should excel here. Traditional Chinese aesthetics attaches great emphasis upon emptiness and blankness, which is exemplified by the use of white space in painting, linguistic allusions within poetry, as well as lingering overtones in music. These are what we call "regarding white as black". The architectural circle often enjoys quoting a passage from Lao Tze's Book of *Daodejing* (or Tao-te Ching), which says: thirty spokes converge at one hub, but the cart's utility results from the nothingness inside; clay

乾隆北京鼓楼街区地图
Map of the Gulou district of Beijing under Qianlong reign

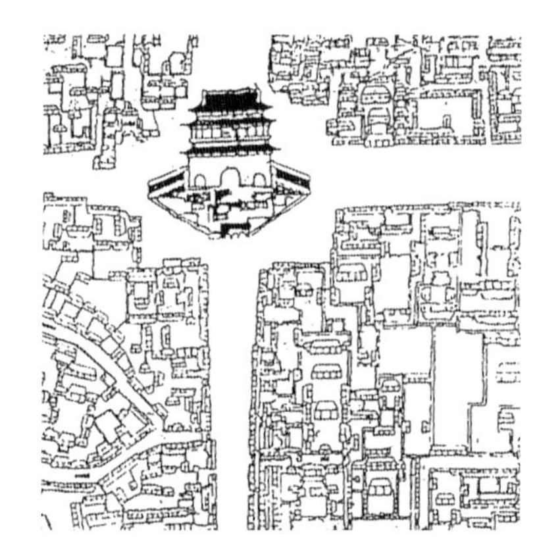

is formed to shape a pot, but the pot's utility results from the nothingness inside; doors and windows are chiseled out to make a dwelling, but the dwelling's utility results from the nothingness inside. This passage explains the dialectical relationship between "something" and "nothing". A similar view can be found in The Writings of Tuang-tzu: emptiness is used to deal with infinity, which explains the magical effect of blankness.

A contrast between G.Norian's 1748 map of Rome's Pantheon district and a map of the Gulou district of Beijing under Emperor Qianlong's reign reveals that, in both the West and the East, yards, streets, and squares are important typologies constituting the spatial blankness of a city. The difference is that yards in China are active whereas streets are passive; by contrast, Western yards are passive while streets are active, as will be shown in the example below. Traditionally, few Chinese cities boast squares; the few squares in existence are marketplaces or palace squares formed from an enlargement or extension of yards. For instance, in Chang'an's ancient times, there was both an eastern and a western market, both of which were surrounded by city walls (referred to as "huan" or " 闤 " in classical Chinese). The walls had doors (referred to as "hui" or " 闠 " in classical Chinese) for the entrance and exit of traders. The Chengtian Gate (which is called Tiananmen nowadays) of the Ming and Qing Dynasties had a T-shaped palace square in front of it, with walls erected on the eastern, western, and southern sides. Due south, there was the Daming Gate; to the east was the Chang'an Left Gate and the west the Chang'an Right Gate. In contrast, Western city squares are mostly an extension or enlargement of streets. The Piazza San Marco of Venice—"the drawing room of Europe"—faces the sea on one side and connects with the city through numerous streets radiating from it on the other sides. Such drawing room style squares dot the ancient cities of Italy and play an important part in everyday life. Yards in western residential houses are used mainly for lighting, airing laundry, or even for storing odds and ends. Yard walls which face the street are delicately decorated while those which face inward are decorated more plainly, unlike in Chinese courtyards where inward facing walls are more delicately decorated.

The above east-west comparison is an exploration on "positive yard space"—an important type of traditional Chinese architectural space. As a distinct spatial type, it is of value even for today's architectural designs. This is especially so nowadays when grass lawns (for viewing only) and squares (for standing only) have become more and more popular in cities. Do not such attempts bear more meaningfulness with them? The core of positive yard space lies in its "positiveness", no matter whether the yard is an open square, enclosed, or covered.

· *Artistic Conception and Montage*

The tradition of Chinese art puts an emphasis upon the formation of artistic conception. Wang Changling, a poet of the ancient Tang Dynasty, points out, in his essay on the composition of poetry, that a poem has three layers of conception: the physical, the emotional, and the artistic. Su Shi, a Song Dynasty poet, while

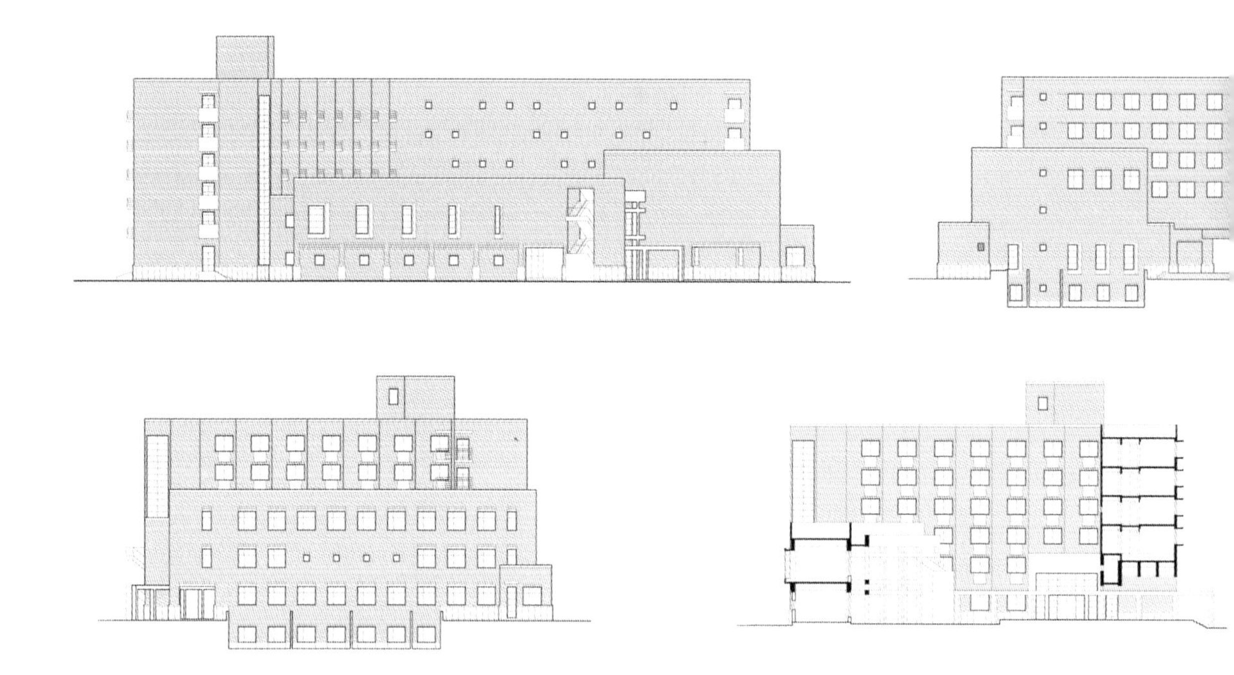

commenting on Wang Wei's poem and painting about the mist and rain of Lantian, says, "I can see a pictorial scene in Wang's poem and I can extract a poem from his painting". It means that emotion is embedded in visualization of the poem and is implied by viewing of the painting. A skillfully produced work of art without an artistic conception is deemed trivial and disregarded in traditional Chinese aesthetics. The Chinese people traditionally had a better understanding of perception than their western counterparts did. The Chinese used marble as hanging screens and communicated with each other by drawing on its stony texture ages ago. In fact, it was nearly 1500 years after Gu Kaizhi of the ancient Jin Dynasty brought up the idea of "integrating emotions into a painting to get a desired effect" that Hegel proposed "conveying spirit through form".

Artistic conception, as an element of traditional Chinese aesthetics, is given importance even in architecture where skills and techniques are invariably applied, though this is especially evident in the design of classical gardens. To be specific, the formation of artistic conception has the four features to be explained below. First, an emphasis on time. In considering the fourth dimension of architectural creation, Chinese traditional architecture has its distinct features. Time is an important factor for creating spatial artistic conception, whether in the imperial Forbidden City or in the Temple of Heaven or in the private pleasure gardens of the south. Second, downplay single buildings. In producing the three dimensions of an architectural work, novelty is not pursued.

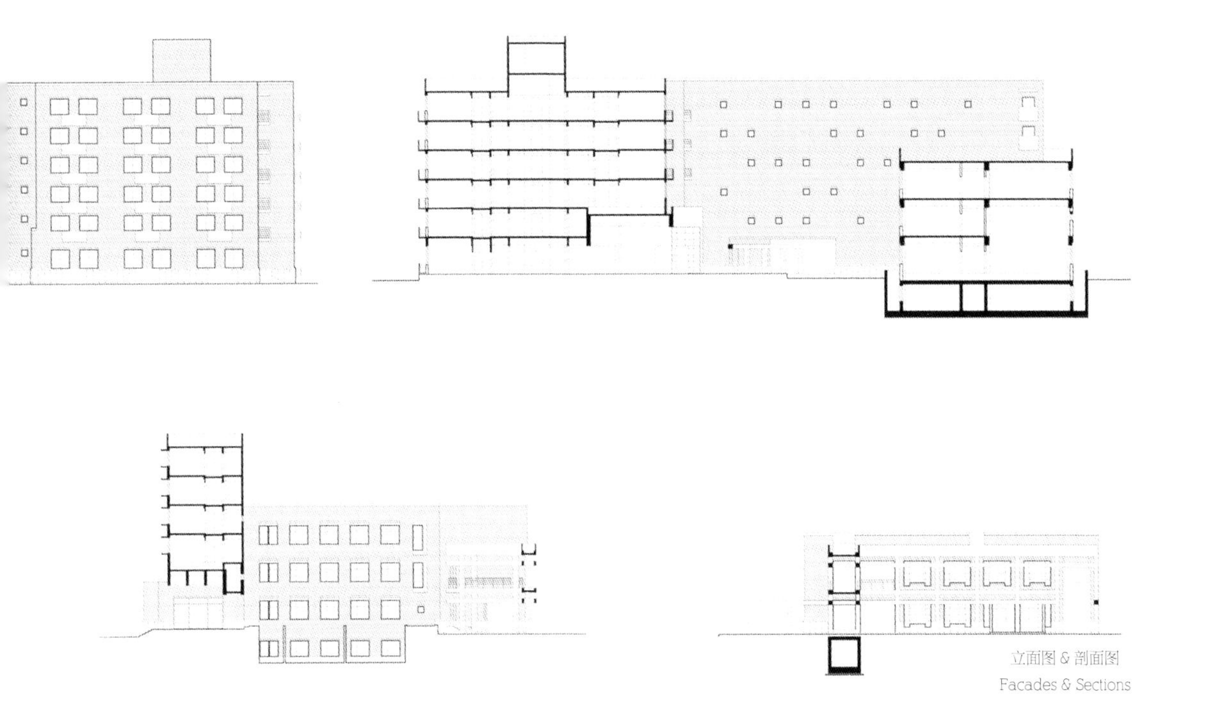

立面图 & 剖面图
Facades & Sections

Even such single buildings as the Taihe Palace and the Qinian Palace are less extravagant than western-style buildings. Third, an integration of the real and the unreal. Shen Fu, in *Fusheng Liuji* (Six Chapters of a Floating Life), says, "One should try to show the small in the big, and the big in the small, and provide for the real in the unreal and for the unreal in the real. One reveals and conceals alternately, making it sometimes apparent and sometimes hidden. This is not just rhythmic irregularity." Fourth, literary texts used as embellishment. The Chinese are adept at linking architecture—the most physical form of art—with poetry—the most spiritual form of art. A tablet or an antithetical couplet is often the perfect finishing touch to the artistic conception of an architectural work.

Artistic conception does not come from a simplistic combination of different images. Instead, it is more similar to the cinematographic technique of montage. The effect of two shots set side by side is not equivalent to the effect of them montaged. Rather, the newly created image works far beyond the scope of the original images. This is what is called "images beyond images" in Chinese aesthetics.

The artistic conception emphasized in traditional Chinese architecture can still be a source of inspiration for today's architectural design, this is especially so for urban design. While learning from the west, we should bear in mind our own valuable traditions. It is the architects' responsibility to integrate them into today's society. We should put more emphasis on spatial artistic conception rather than on the novelty of form. The increasingly chaotic cityscape and our appetite for monuments and alien forms reveal that artistic conception has long been kept out of mind.

· *Dematerialization and Emotionalization*

With the advent of the information era, a new trend—the dematerialization of architecture—has begun to take shape. This so-called dematerialization refers to a substitution of traditional building materials (such as bricks, stones, and wood) with new materials (such as glass, steel, and metal plates) so as to foster a concept of dematerialization, or rather, a concept of using unconventional materials. Naturally, this brings about an aesthetic change. That is why it is often said that "if you want something new, use metal; if you want something cool, use a glass or metal screen". However, no matter what materials are used, conventional or unconventional, they cannot be the final expression of an architectural work. Lying behind all forms and materials is emotion.

Indian architect Charles Correa, while talking about his own experiences in architectural creation, says architects should use modern technology to represent the mystery and beauty of Oriental culture. A building is deemed vulgar because it has nothing to do with human emotion. Correa himself is a practiced hand at expressing emotions with conventional materials. Those masters using unconventional materials for self-expression have also changed tremendously. The Italian architect Renzo Piano's Tjibaou Cultural Centre in New Caledonia has made do without the machine-like coldness of earlier high-tech buildings, and thusly became a dancer against the background of the blue sky and the sea. He says that the universality of a building in the true sense of the word should be gained through root-seeking, showing gratitude for historical bestowment, and respecting history and culture.

Materialization or dematerialization are not the key of this issue. A French architect who has come to Beijing and visited the Beijing Film Academy's Yifu Film & TV Art Centre has testified that grey bricks can make a nice building all the same.

The above discussion is intended to better acquaint the public with Wang Yi's intentions and aspirations, which were consulted throughout the design of the Film & TV Art Centre. All in all, it is possible to say that what is seen is form and what is felt is emotion, but ultimately the building plays the decisive role.

Due to this project, I have become a frequenter of the Beijing Film Academy. Students incessantly visit the construction site for purposes of internship or photography. Once, a student majoring in photography said that this building has quite a number of excellent angles and settings from which he can shoot. A student of the performing arts told me that the whole setting of the building helps to create a special atmosphere suitable for backdrop and acting. What's more, a vocal music major, in his excitement, clued me in on his plan to shoot a music video there. The fondness of students for a built project is an architect's best reward.

Originally published in *Architecture Journal*, 2000 (10)

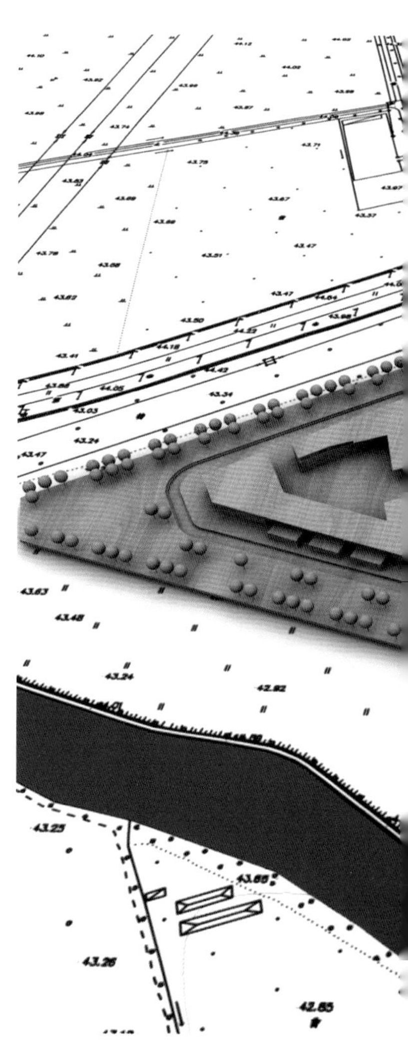

中国纪检监察学院
China College of Supervision

主题：再利用

Theme: Reuse

在中国纪检监察学院建设用地内，遗留有大量原软件培训中心和国际学校的建筑，总建筑面积达5.8万平方米。如何利用这些旧建筑成为新校园规划设计遇到的关键问题。

本方案将旧建筑视为一种资源，将它们巧妙地融入新的建筑群体之中，以适应新的使用要求。新的建筑群体从东到西分为三个组团，一根轴线从东向西贯穿其中。新的道路体系，通过曲或直的变化，化解了地段的不规整以及原有建筑物所造成的

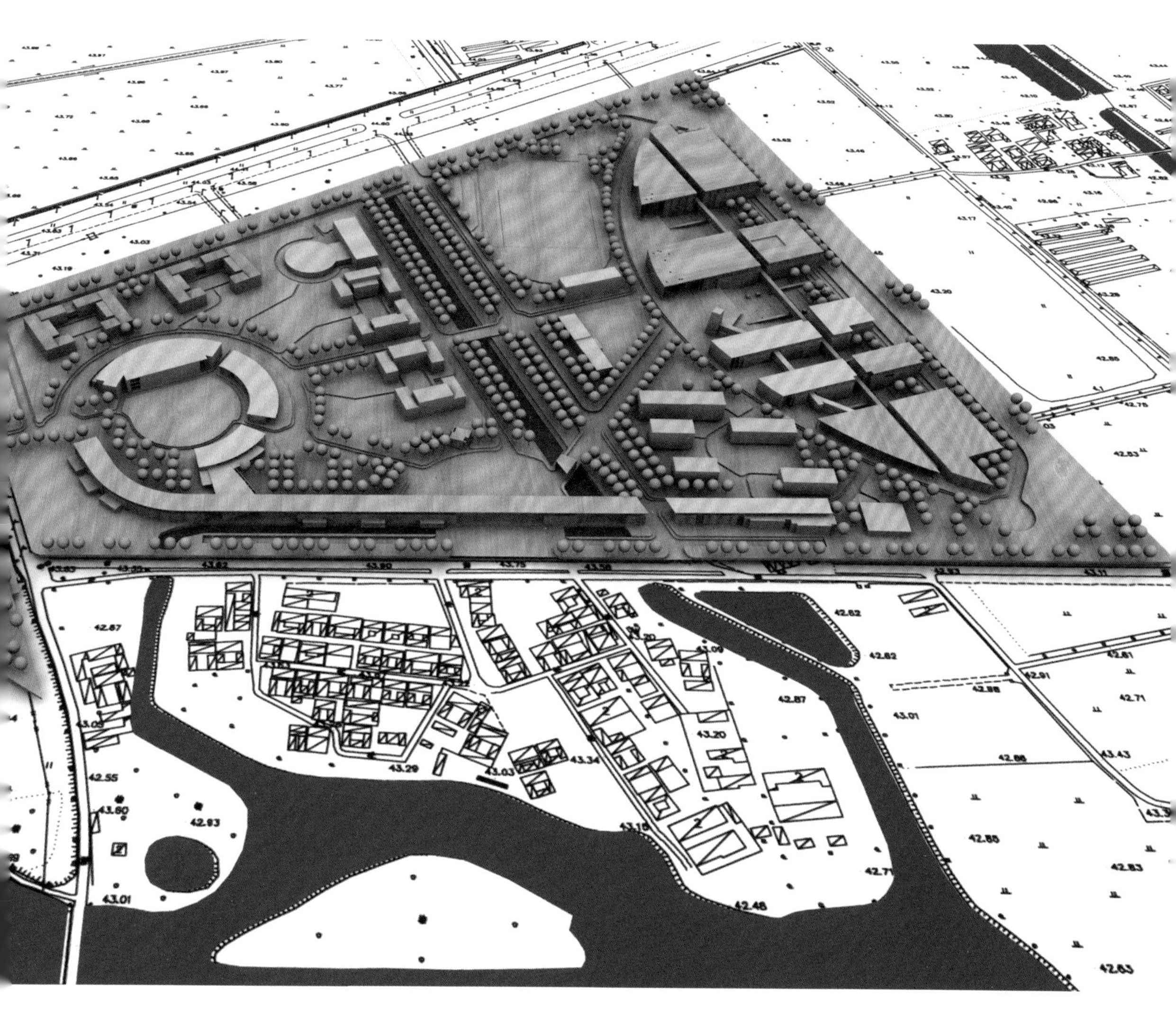

矛盾。除了旧建筑外，本方案还充分利用了地段的自然条件，尽量保留原有水系和植被，把阳光和绿色引进建筑群当中。

利用旧建筑并不意味着新校园只能拘泥于旧有的建筑形式。相反，本方案致力于在空间布局上艺术地处理新旧建筑群的关系，将两者巧妙融合为一体，并呈现出新的艺术形象。另外，本方案不追求建筑单体的雄伟和宏大，而是通过建筑群体来渲染和烘托校园的环境。在建筑群体当中安排通廊，方便不同建筑之间的联系。在建筑立面上力求端庄、简洁，避免乖戾夸张的造型，以体现院校类建筑的特性。

The China College of Supervision inherited a massive amount of buildings from the software-training centre and foreign institutes, netting a total surface area of 58,000 sq. metres. Architects encounter the dilemma when applying

现存旧建筑
Exisiting old structures

新建筑与旧建筑融为一体
New buildings intergrated with old ones

the existing buildings in a new campus planning.

This project views existing buildings as resources; brilliantly fuse them between new buildings to adopt to their new use. Three groups of brand new buildings are linked horizontally from east to west. A new roadmap neutralizes uneven sectors and contradictions from existing buildings through different

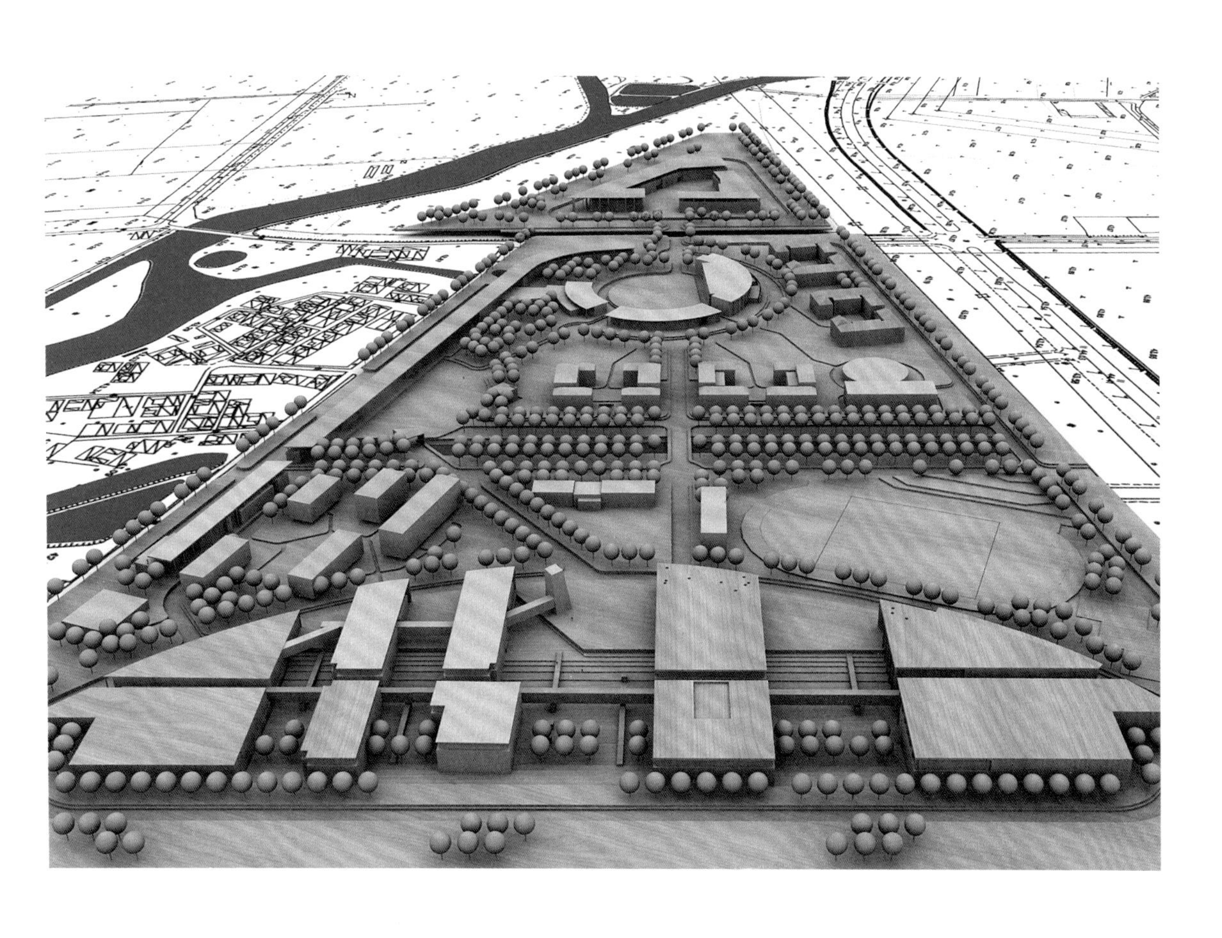

straights and bends. This project utilizes the most of the soil's natural qualities. Besides existing buildings, retaining maximum amount of hydrography and vegetation to lead sunlight and green inside the structures. Using existing buildings does not mean this project is limited to traditional architecture; in contrast, this project devotes itself to fusing new and existing buildings. Therefore, a new artistic layout emerges from the fusion. The architecture of China College of Supervision does not pursue the magnificence of a single building; the focus is to colour and shade the campus with groups of buildings. Promenades connect building groups for convenient access. In the aspect of architectural elevation, campus buildings are to be dignified, concise, and not exaggerating.

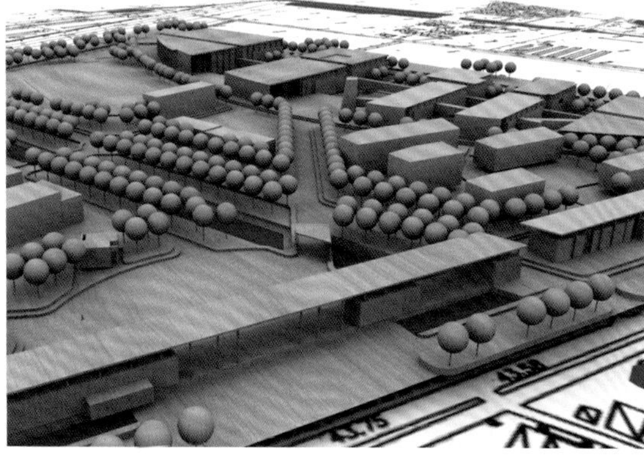

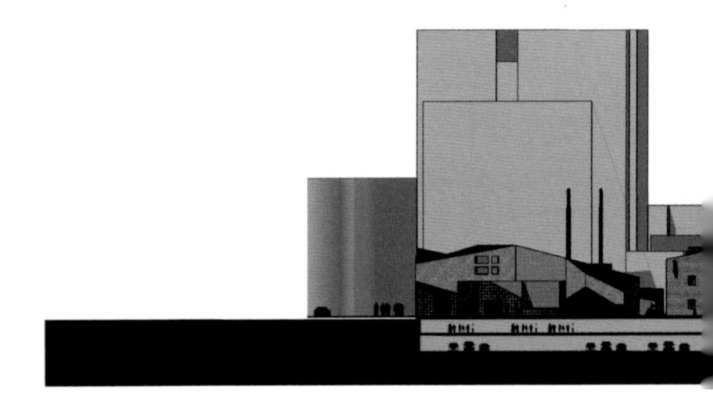

大小马站历史街区

Daxiao Mazhan Historical District

主题：文脉再造

Theme: Context Reconstructed

书院，是古代一种有别于官学和私塾的教育机构。广州书院群最早始建于明朝，到清代中后期，曾出现过全国罕见的书院群，达数百间之多。广州书院对岭南文化的发展起到非常重要的作用，但在近几十年的城市快速发展过程中，被成片拆毁，没有拆毁的也由于年久失修，变得破烂不堪。位于大小马站的书院群就是这仅存的部分。

基于现状条件，大小马站书院群已经无法形成完全独立的街区，本方案也不以单纯的保护和恢复

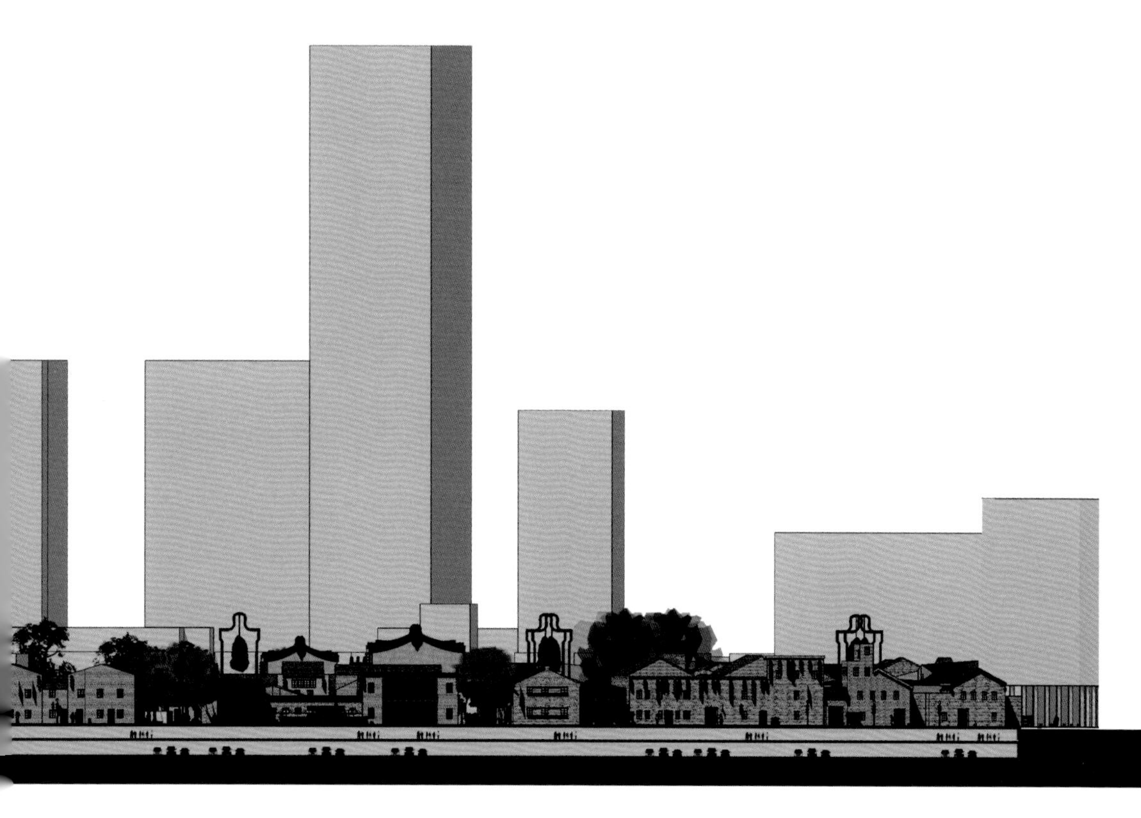

古书院群为目的，而是根据当地经济文化的发展状况，适当加入新的、时尚的元素，对历史文脉进行再造，以吸引和适应年轻一代的需求。

依照国内外成功的经验，对旧建筑的合理使用是延续建筑寿命的最好方法，博物馆式的保护长久不了。本规划设计方案对旧建筑做了适当改造，置入合理的使用功能，激活院落场所，将私人空间转化为公共空间。另外，对地下空间进行了适当开放和利用，以补充历史街区空间容量的不足，并置入现代化建筑设备。

Colleges are a kind of academic institutes separated from government and private schools in ancient China. The first colleges in Guangzhou were built during the Ming Dynasty; a group of colleges emerged at the times of mid to late Qing Dynasty, which was very uncommon in the whole China at time. Colleges in Guang-

zhou area took significant role in the development of *Lingnan* (south of the Five Ridges) culture, but the majority of them were demolished from the rapid developments of the cities in recent years. Remaining colleges are dilapidated due to poor or no maintenance. The colleges in Daxiao Mazhan Historical District are one remaining that survived.

The colleges in Daxiao Mazhan Historical District are not capable of forming an independent district in its current state. The project's goal is to forge a historical context base on local economic and cultural growth. Instead of purely rebuilding and protecting the colleges, innovative and fashionable elements were suitably added to attract the youth and satisfy their needs.

Referencing international achievements, the best way chose to extend the life of architec-

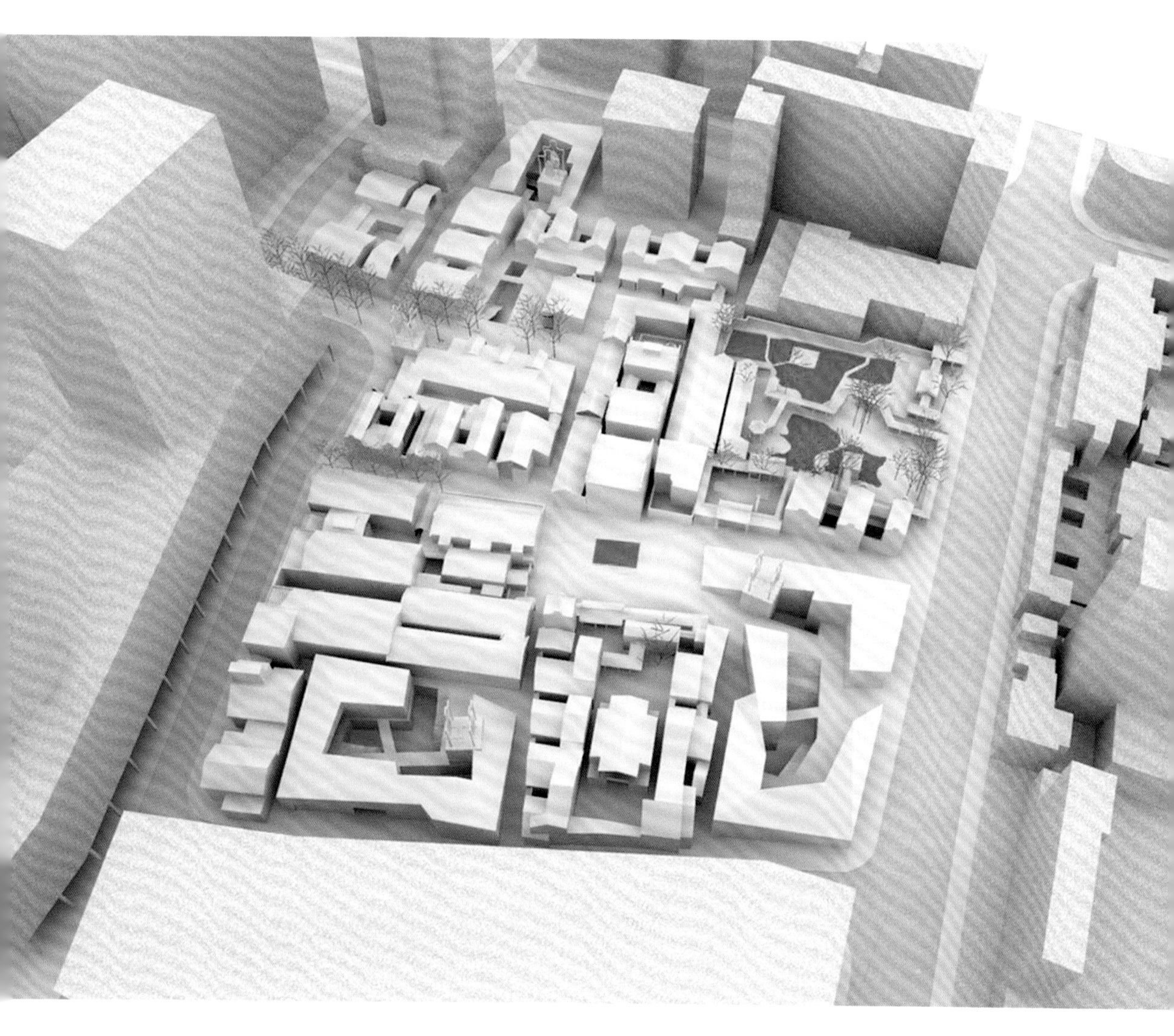

tures is fair usage; museum-approach treatment will never last in architecture. The project properly renovates aged buildings, adds proper functions, stimulates the courtyards and concourses, and transforms personal spaces into public spaces. In addition, underground spaces were properly opened and used with added modern architectural facilities, to supply the historical district's lacking capacity.

总图 Plan

周边街区功能

Functions of the blocks around

1. 分区 Zoning

2. 建筑等级 Architectural hierachy

3. 绿化空间 Green space

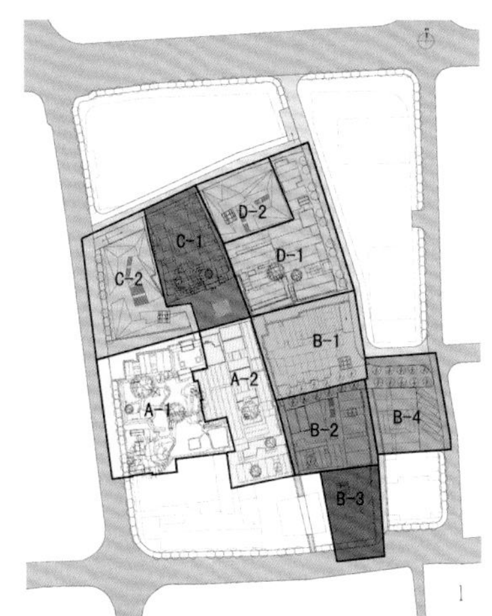
D-2
C-1
D-1
C-2
B-1
A-2
A-1
B-2
B-4
B-3
1

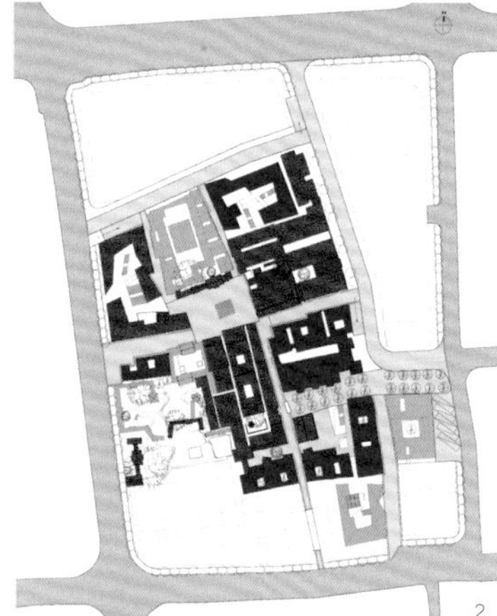
2

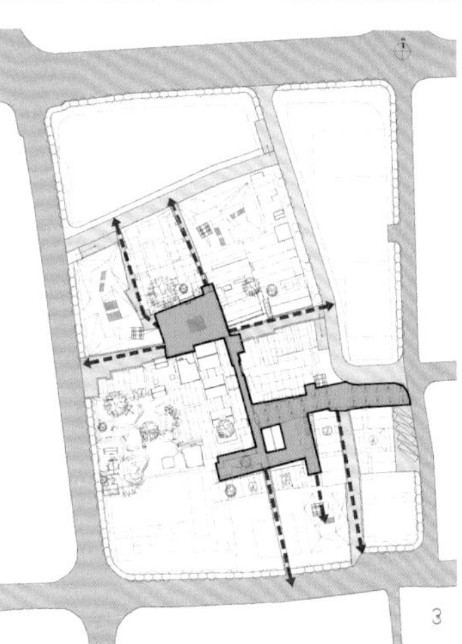
3

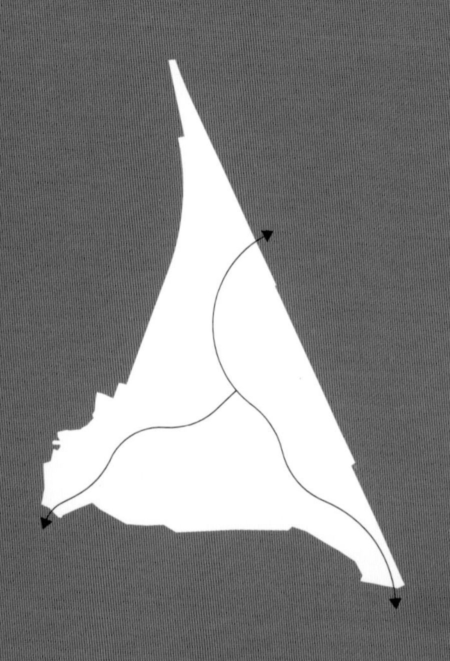

窑埠古镇
Yaobu Ancient Town

主题：船如满月

Theme: A Boat as Full Moon

窑埠古镇历史上是柳州砖窑集中的地方。烧制的砖瓦需要船只运输出去，所以又形成码头。砖窑、码头、小巷、旧屋，昔日窑埠古镇独特的地域风景随着城市的发展而渐渐消退。拟建中的窑埠古镇将成为集传统特色和现代风貌为一体的城市景观节点，承担旅游、休闲、购物、居住等多项城市功能。

本方案为窑埠古镇勾勒出一个非常明晰的空间结构——酒店依水，山寨傍山，住区临路，古街巷居中。这个空间结构，从景观的角度看分为三个层

次。第一层次是酒店、游廊和码头，在大尺度上与周围的山水和城市相呼应，成为百里柳江之山水城市画面的重要片段；第二层次是古街巷和山寨，在较小尺度上构成窑埠古镇的街道及广场肌理；第三层次为主题花园，在更小的尺度上构成优美的园林景观。

位于地段中心的民俗博物馆是窑埠古镇传统文脉的集中体现，它在造型上强调两个地域元素——砖窑的烟囱和码头的渡船，并将两者巧妙地融为一体，让人联想起装满砖瓦的渡船。虽然没有直接仿照传统建筑式样，它仍然很好地体现了窑埠古镇的传统意境。位于水边的酒店、游廊和码头在造型上强调现代感，使窑埠古镇不仅仅局限于传统的意义，也体现现代的活力。

Yaobu Ancient Town is the centre of Liuzhou brick kiln throughout history. Piers were built for sea freighting baked bricks and earthen

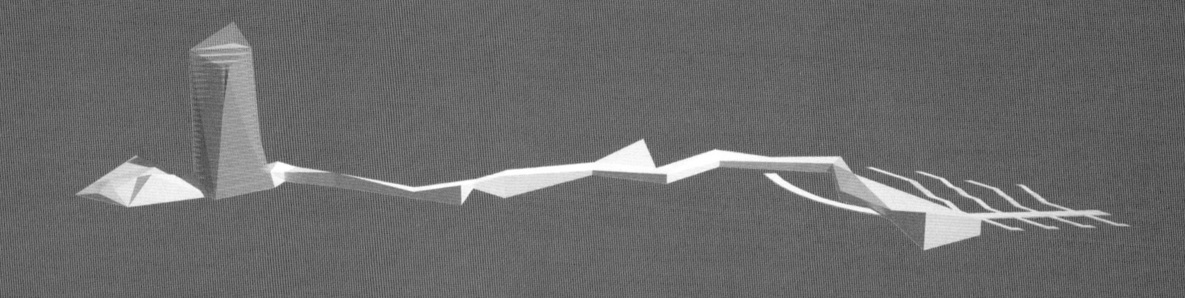

tiles. Landscapes unique to Yaobu Ancient Town such as Brick kilns, piers, alleys, and old residences gradually vanished due to city developments.

The proposed replicated Yaobu Ancient Town will become the city's charm with tradition features and contemporary scenes, shouldering city functions such as tourism, leisure, commerce, and residence. The project sketches a very clear contour of Yaobu Ancient Town replica's structure with hotels close to water, villages close to mountain, residences close to roads, and alleys in the centre. The structure is divided into three levels in the aspect of landscaping. In the first level, hotels, galleries, and piers echo with the city, mountain, and water nearby in a large scale; it will become an essential scene in the city landscape of the Liujiang River area.

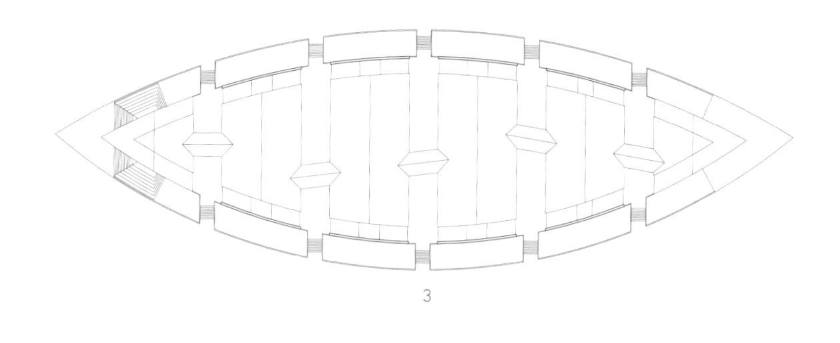

3

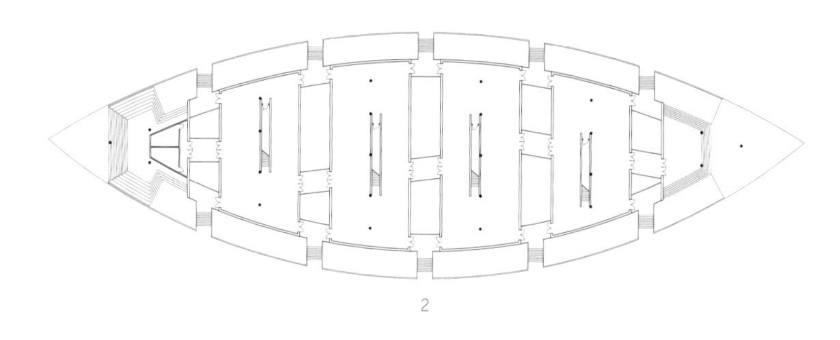

2

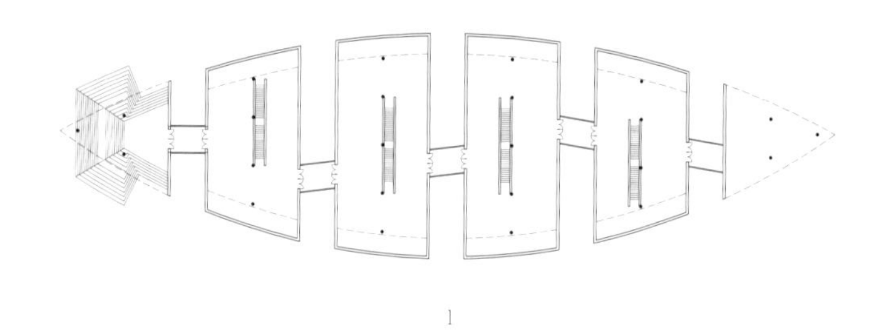

1

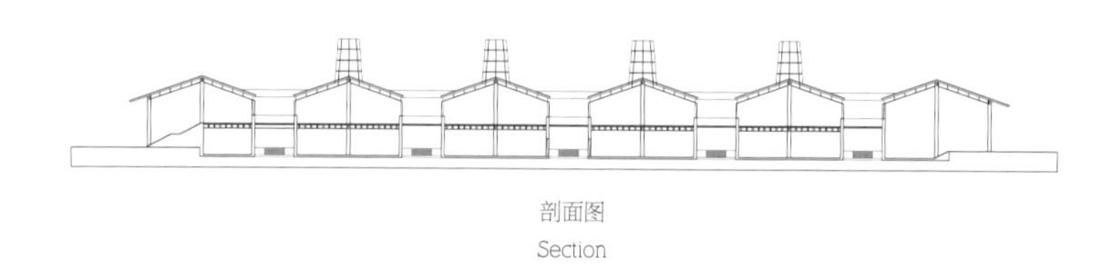

剖面图
Section

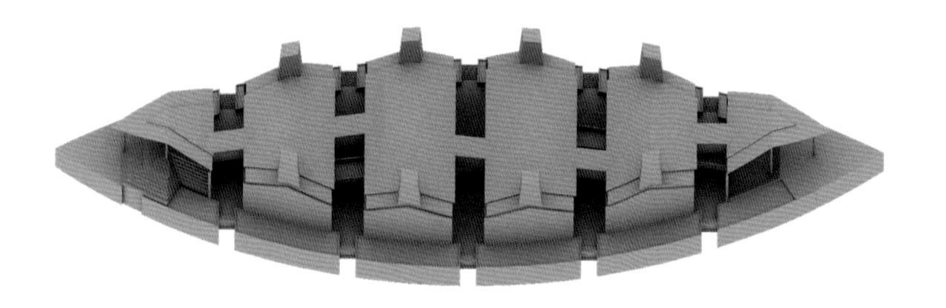

In the second level, alleys and villages construct the building blocks of the streets and plazas of the Yaobu Ancient Town in a smaller scale. In the third level, the theme gardens forms an exquisite view on a even smaller scale.

The National Museum, located in the centre of the district, is the concentrated reflection of Yaobu Ancient Town's traditional context; its appearance emphasizes two local elements - brick kilns and ferries of the piers. The ingenious fusion of the elements draws a picture of ferries carrying bricks and tiles in the people's minds. Although the National Museum is not a direct replication of traditional architectures, it clearly reflects the traditional image of Yaobu Ancient Town. The appearance of hotels, galleries, and piers emphasizes on modern sense, allowing Yaobu Ancient Town to show contemporary vigor on top of the traditional sense.

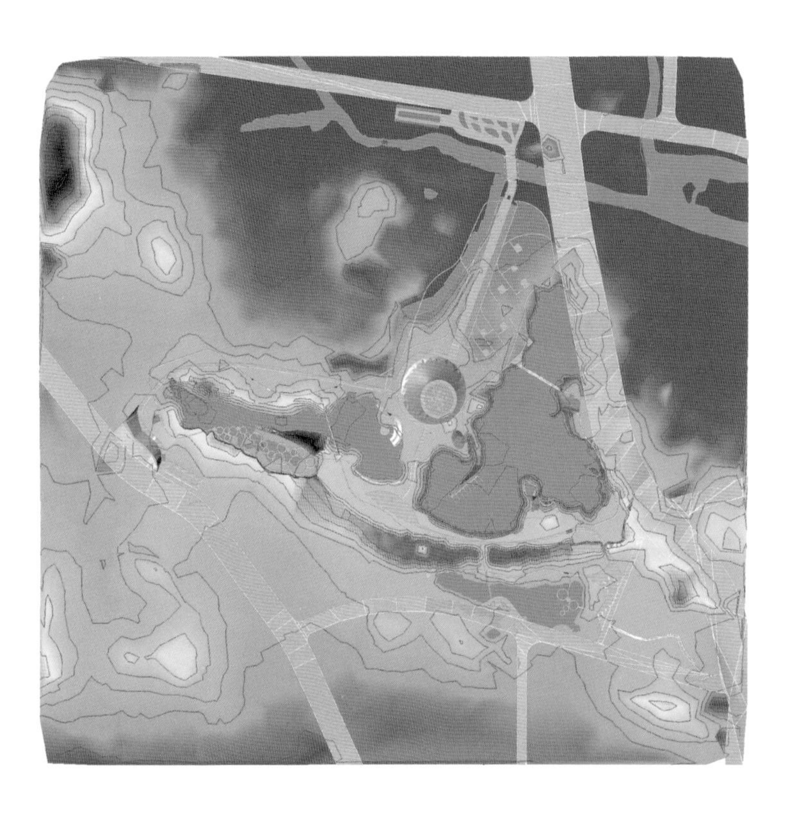

青龙山生态公园

Qinglong Mountain Eco-park

主题：修复自然

Theme: Repairing the Nature

这里原本是城市中一座秀美青山，多年来用作采石场，山体几乎挖光。山只存在名字之中，而留下的是贫瘠的乱石和积水大坑，成为城市的一道伤疤，乏人问津。宜兴市丁蜀镇以紫砂壶闻名于世，青龙山生态公园位于其中心地段。紫砂壶为宜兴带来荣誉，也给其环境造成破坏和污染。在生态理念逐渐成为世人共识的今天，如何减轻污染，修复被破坏的自然成为城市面临的问题。

对于这片被破坏、被遗忘的地块，本方案并不

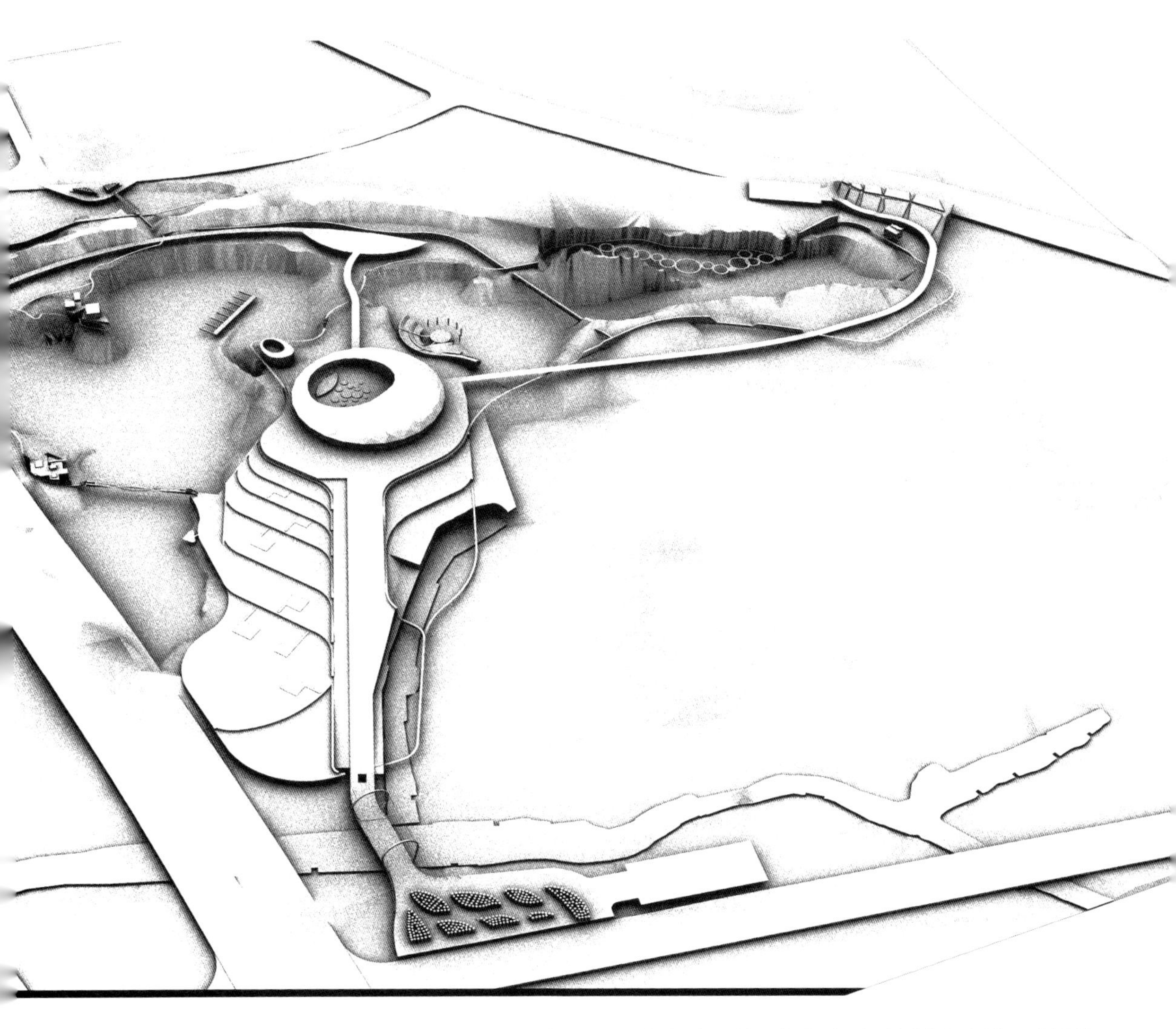

急于完全抹去采石场的历史，重新人为制造一个自然，而是采取修复自然的做法，接受和利用采石场遗留的地形地貌，整合破碎凌乱的场地，恢复绿化植被系统，并适当展现采石场遗留的地貌特征和历史信息。同时，根据城市的发展需求，本方案为青龙山生态公园置入多样的城市功能，向城市居民提供一处具有休闲、度假、娱乐、科普等多功能、设施配套完善、开放友好的城市公共场所。

生态公园并不仅仅意味着树木和绿化。把握好修复的尺度，在不破坏生态特质的前提下，景观和建筑也要不失特色和时代感。

This area was originally an elegant green mountain in the city, but the body of the mountain was almost dug away from years of use as a quarry. The only remaining of the mountain other than in the texts is the barren land full

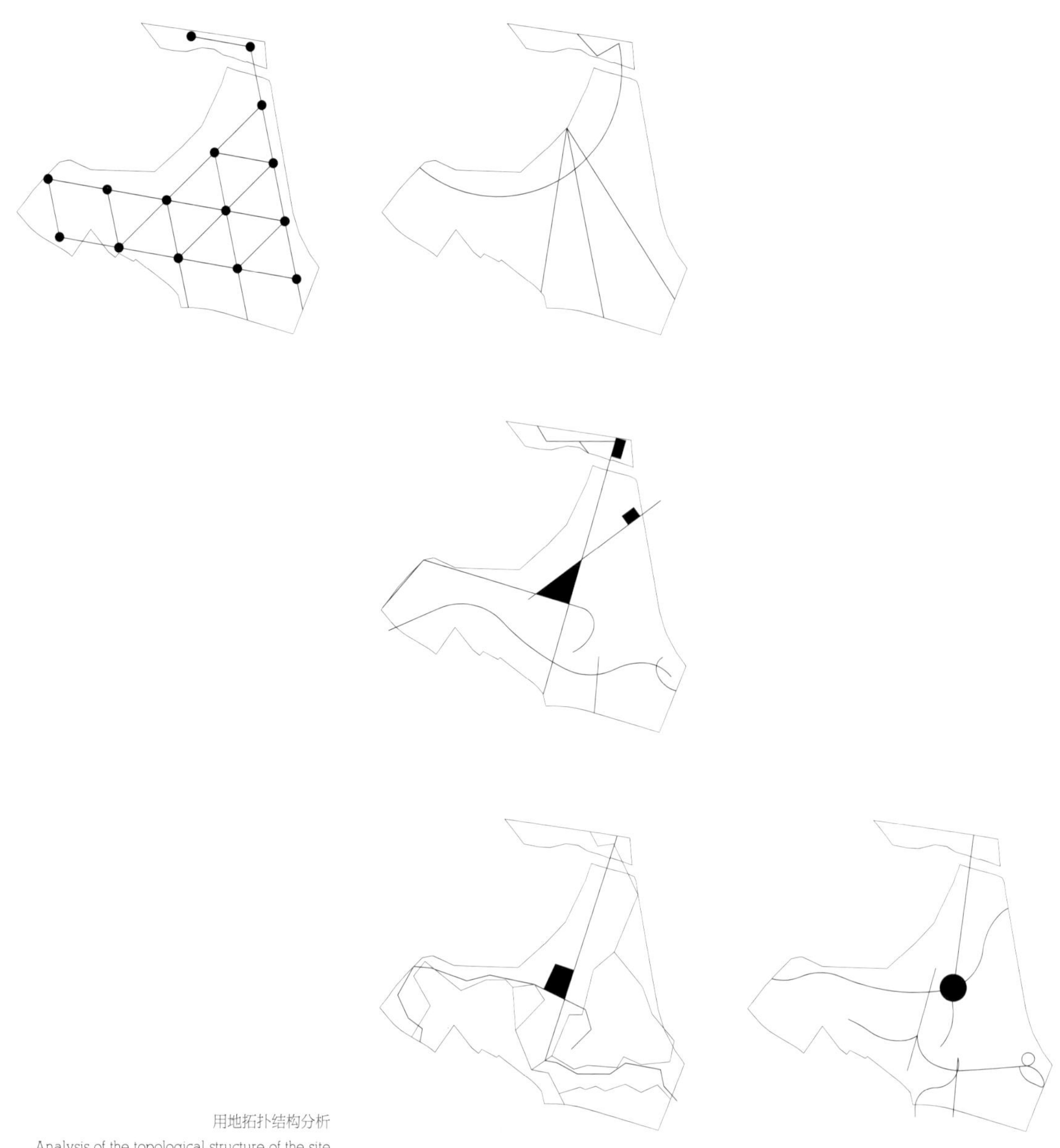

用地拓扑结构分析
Analysis of the topological structure of the site

of rocks and water pits; this becomes a scar of the city with no one's interest. The Dingshu town in Yixing City, where the Qinglong Mountain Eco-park occupies the centre of, produces world-renowned pottery named Zishahu. Zishahu brings honour to Yixing City, but at the same time it brings harm and pollution to its environment. In today's eco-conscious world, cities face the question of reducing pollution and repairing the destroyed nature environment.

The project's goal is neither to rapidly cast away the history of the quarry on this destroyed and forgotten land, nor artificially replicate its natural environment. Instead, the goal of the project is to naturally restore the environment, utilize the characteristics left from the quarry and rearrange the crushed land

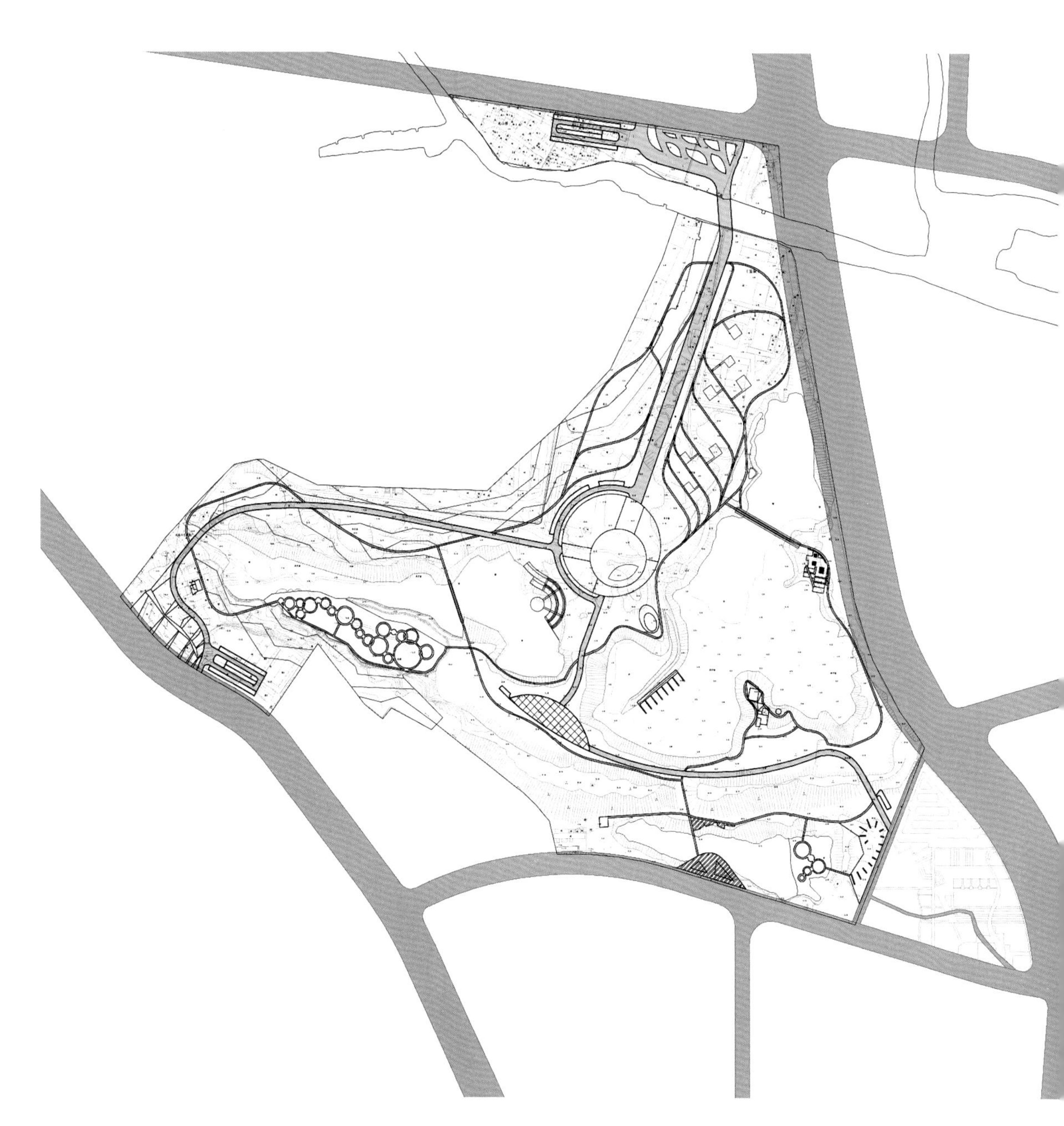

by replanting. The overall focus is to show the characteristics and historical messages inherited from the quarry in a suitable way. According to city planning, the project adds multiple city functions to the architecture at the same time, providing citizens with a civilized public space completely assorted leisure, vacation, and entertainment.

The principle of Ecology Park is not limited to planting and green zoning. Under the principle of protecting the eco-system, landscapes and architectures should aim to be special and contemporary while being on the right scale of the restoration.

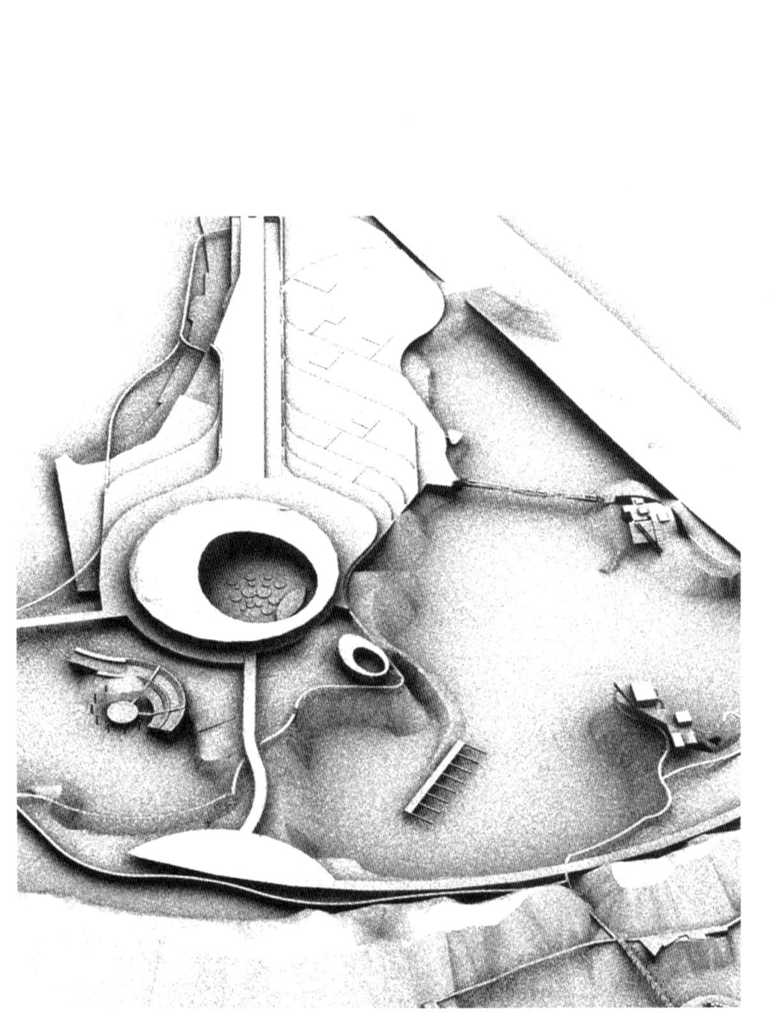

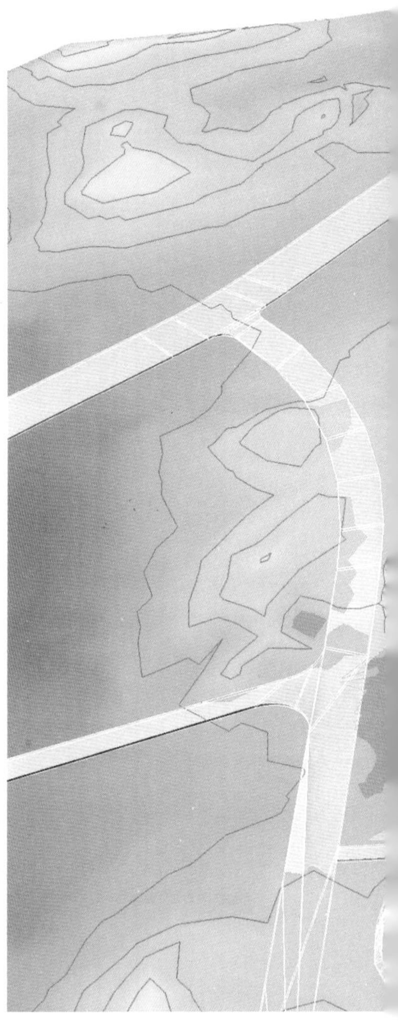

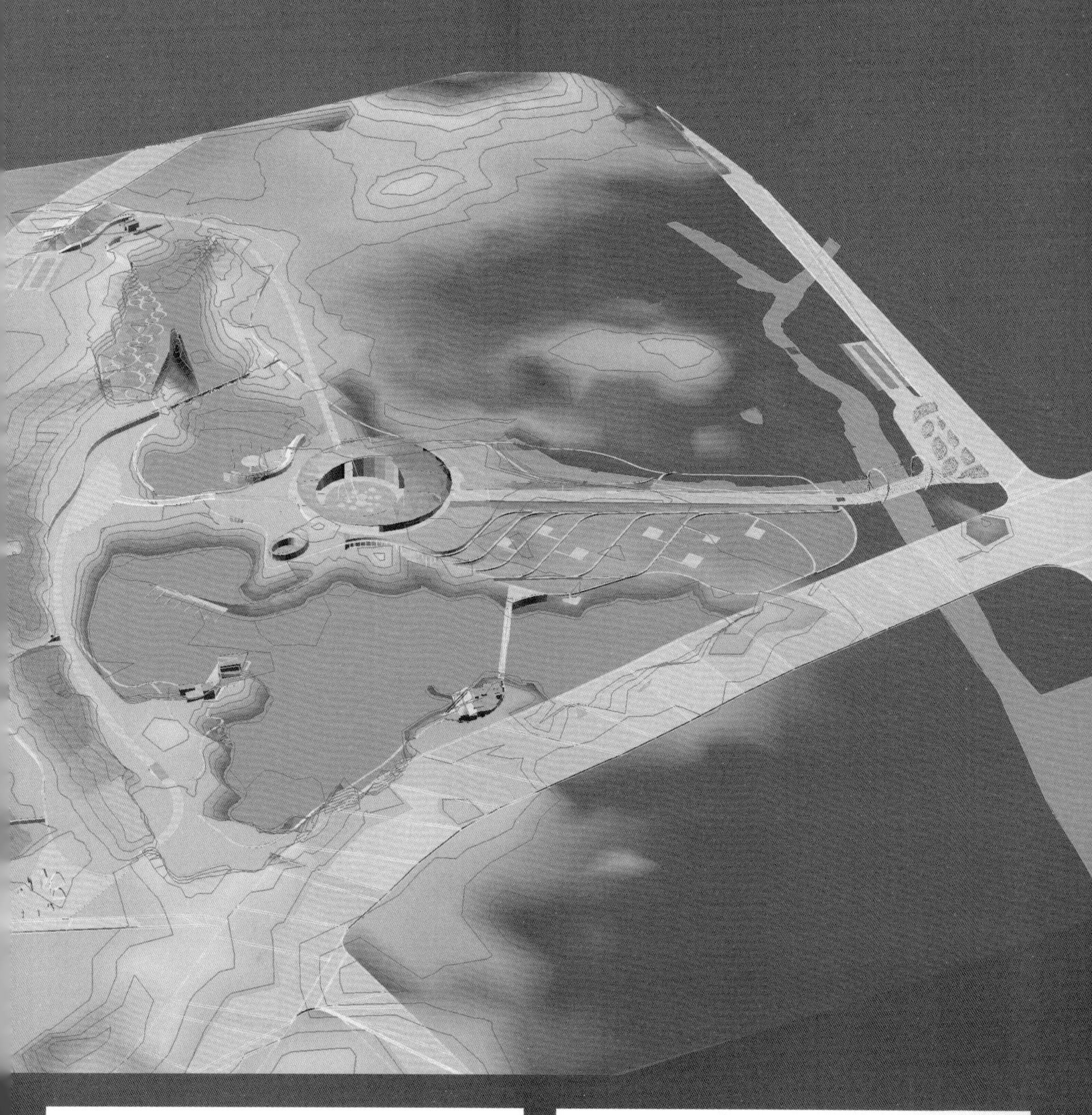

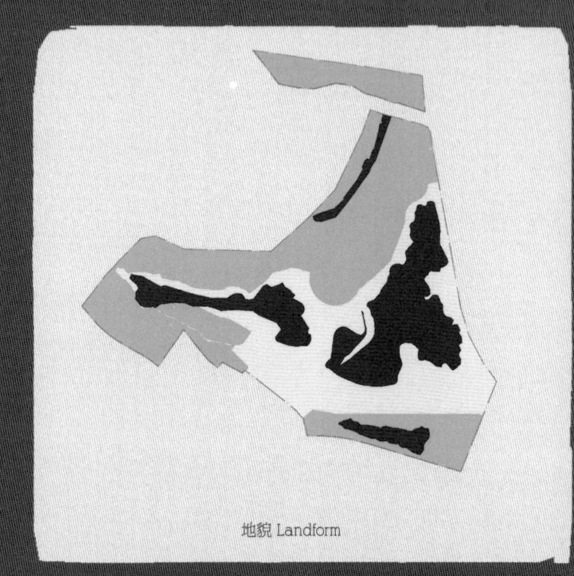
地貌 Landform

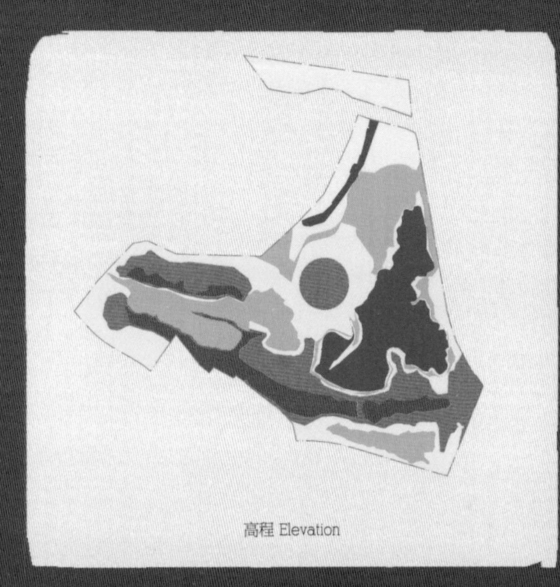
高程 Elevation

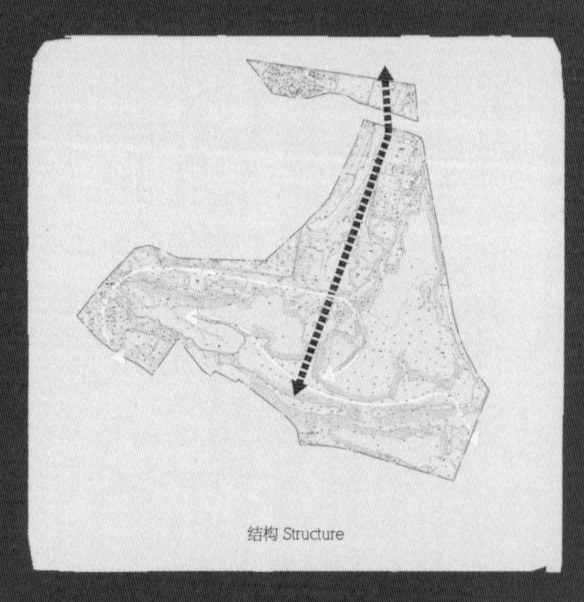
结构 Structure

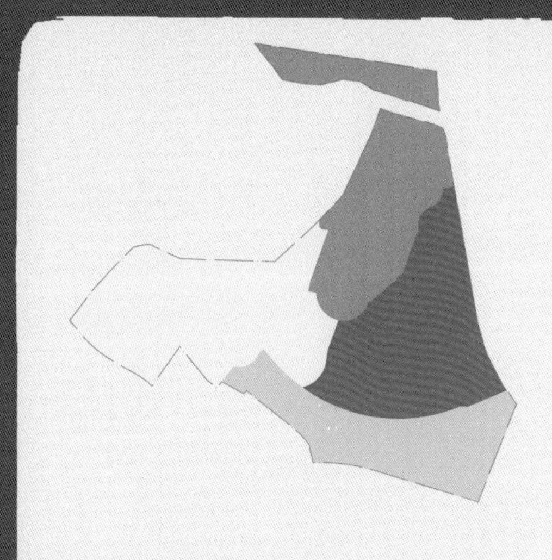
分区 Zoning

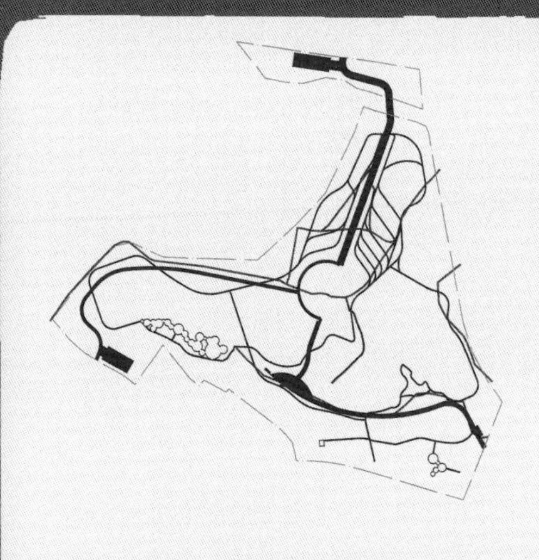
新旧道路 New & old roads

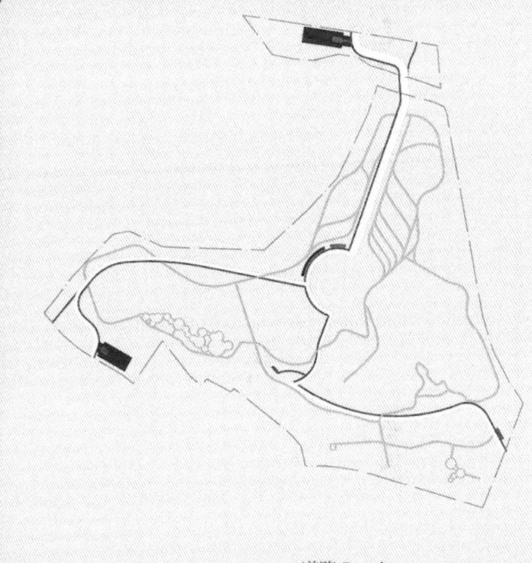
道路 Road

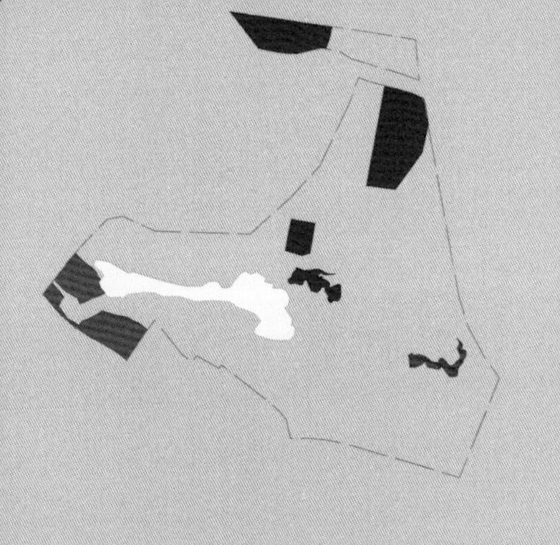
修复手段 Repairing methods

本土地景 LOCAL LANDMARK
文脉更新 CONTEXT RECONSTRUCTED
本土建构 LOCAL TECTONICS
城市空间 URBAN SPACE
本土再造 LOCAL UPCYCLING

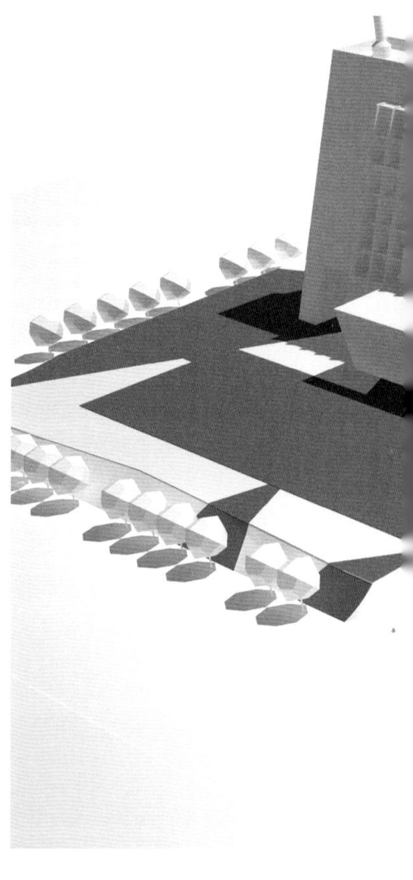

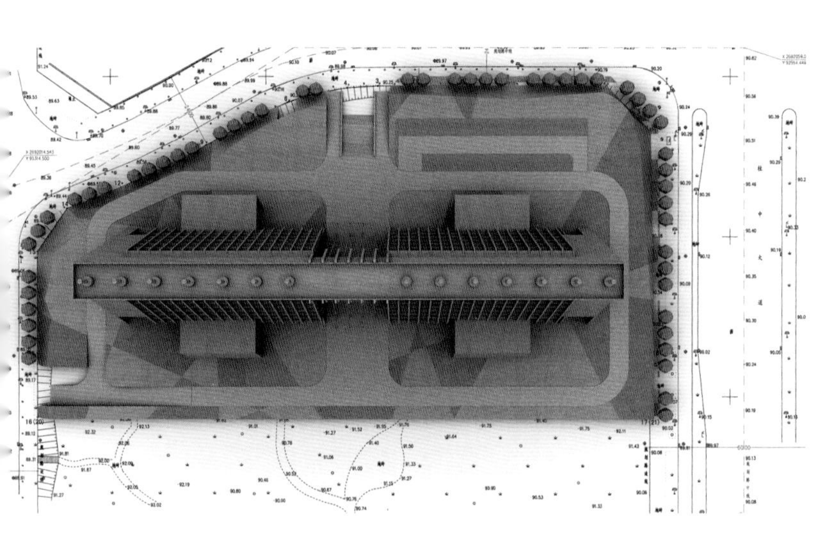

文昌东楼

Wenchang East Building

主题：适宜生态技术

Theme: Proper Ecology Technologies

文昌东楼位于柳州市河东新区。本方案从总体选型到细节设计上均体现当地的气候特点和要求。将建筑中庭空间设计成下宽上窄的形状，并在顶部设置风塔，有利于将室内热空气迅速排出。将建筑底层架空，有利于地表冷空气流入，并为地下车库提供自然采光。建筑外墙呈倾斜面，但80%以上的窗户为垂直安置，窗前种植绿化，形成绿化墙面。本方案秉承经济合理的原则，选择适宜的生态技术，使生态理念不是停留在口号上，而是落实到实处。

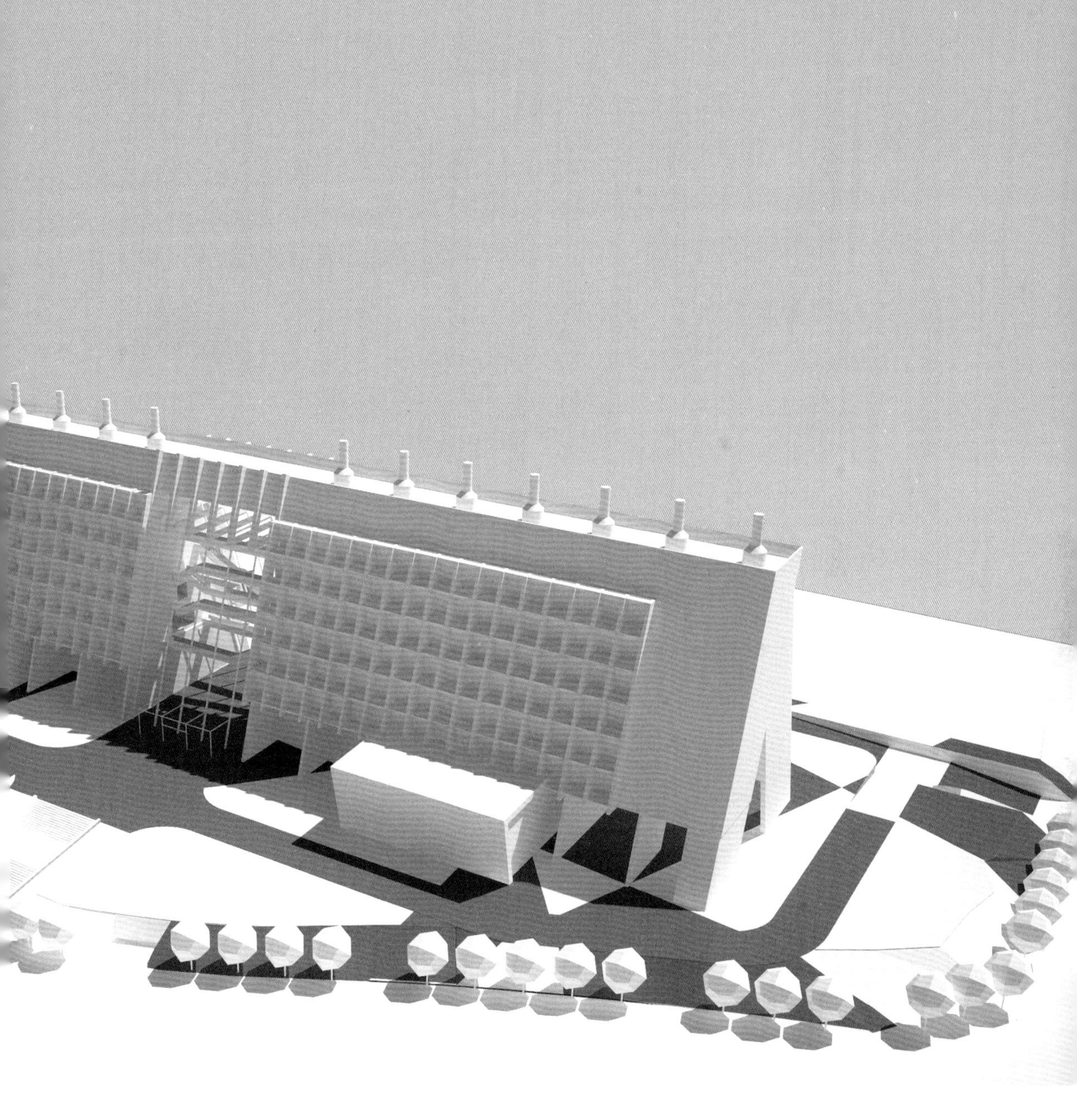

Wenchang East Building is located in the New Hedong district of Liuzhou City. From overall modeling to detailed designing, the project makes great effort to reflect the needs and characteristics of the local climate. A wind tower is built on top of the pear-shaped atrium to benefit the rapid venting of hot air. The building is built without foundations to provide cool air and natural sunlight to the bottom part of the building. The exterior walls are built with an angle while more that 80% of the windows are vertical, providing space to plant and achieve a green wall. The project adheres its principle to be financially rational, chooses the proper ecology technologies to carry out the eco-concept. Eco-consciousness now exists outside of a slogan because of this project.

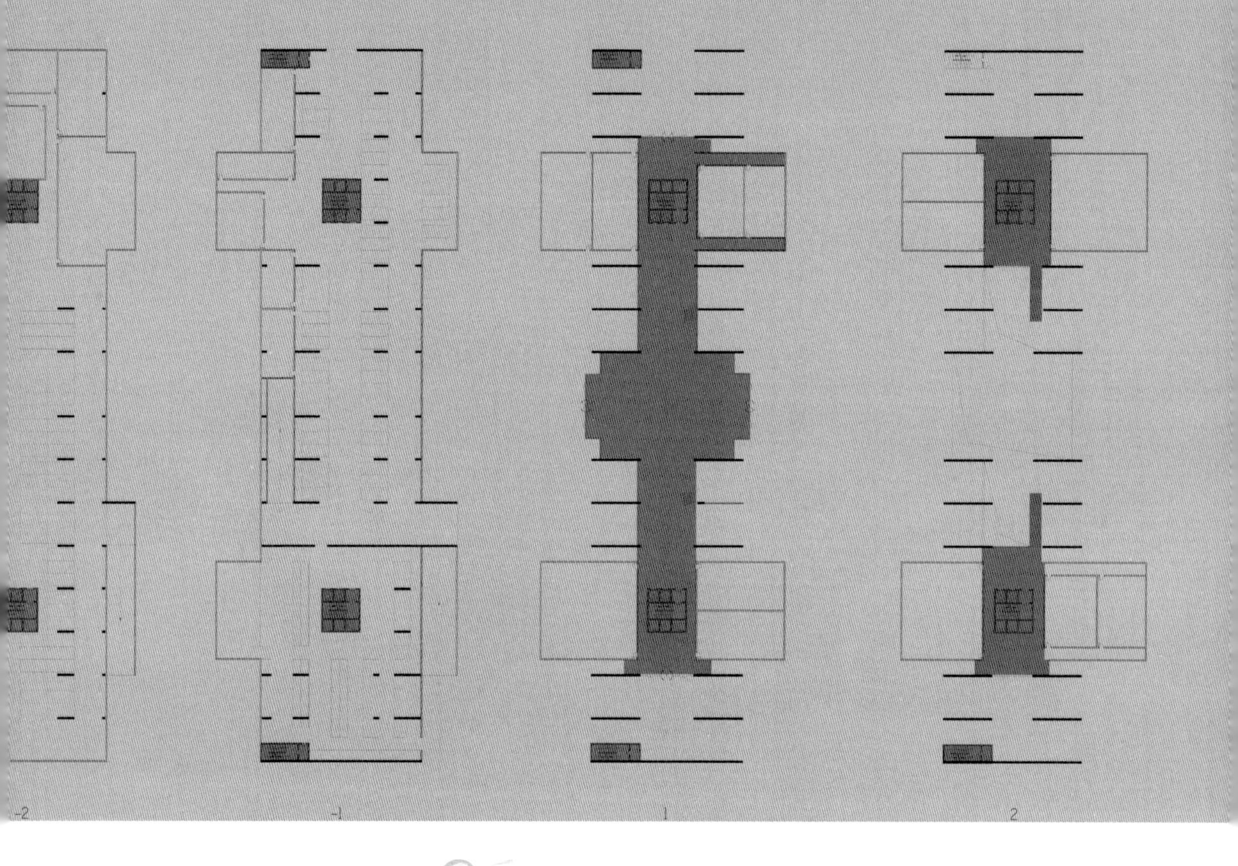

-2
-1
1
2

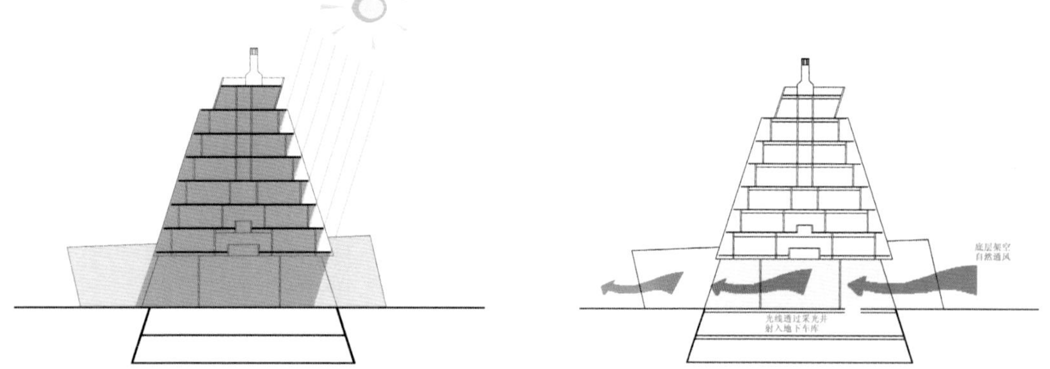

底层架空
自然通风
光线透过采光井
射入地下车库

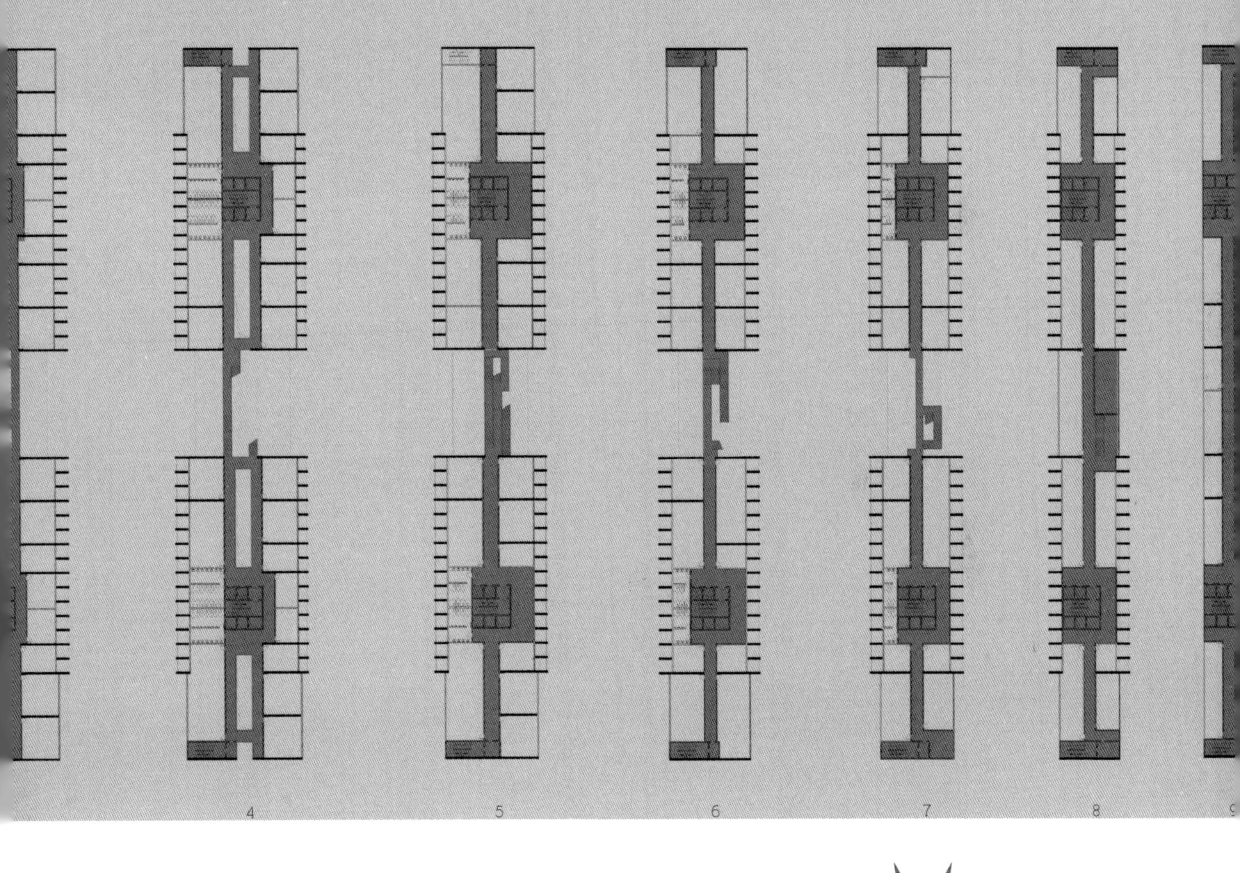
4
5
6
7
8

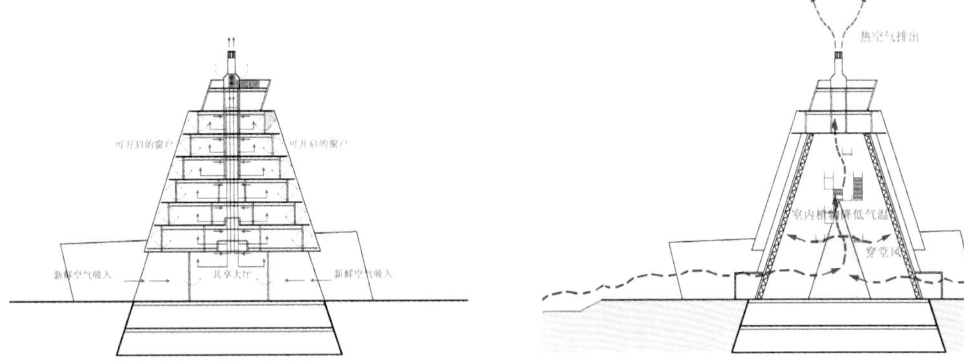
可开启的窗户
可开启的窗户
新鲜空气吸入
共享大厅
新鲜空气吸入
热空气排出
室内植物降低气温
植被降低气温
穿堂风

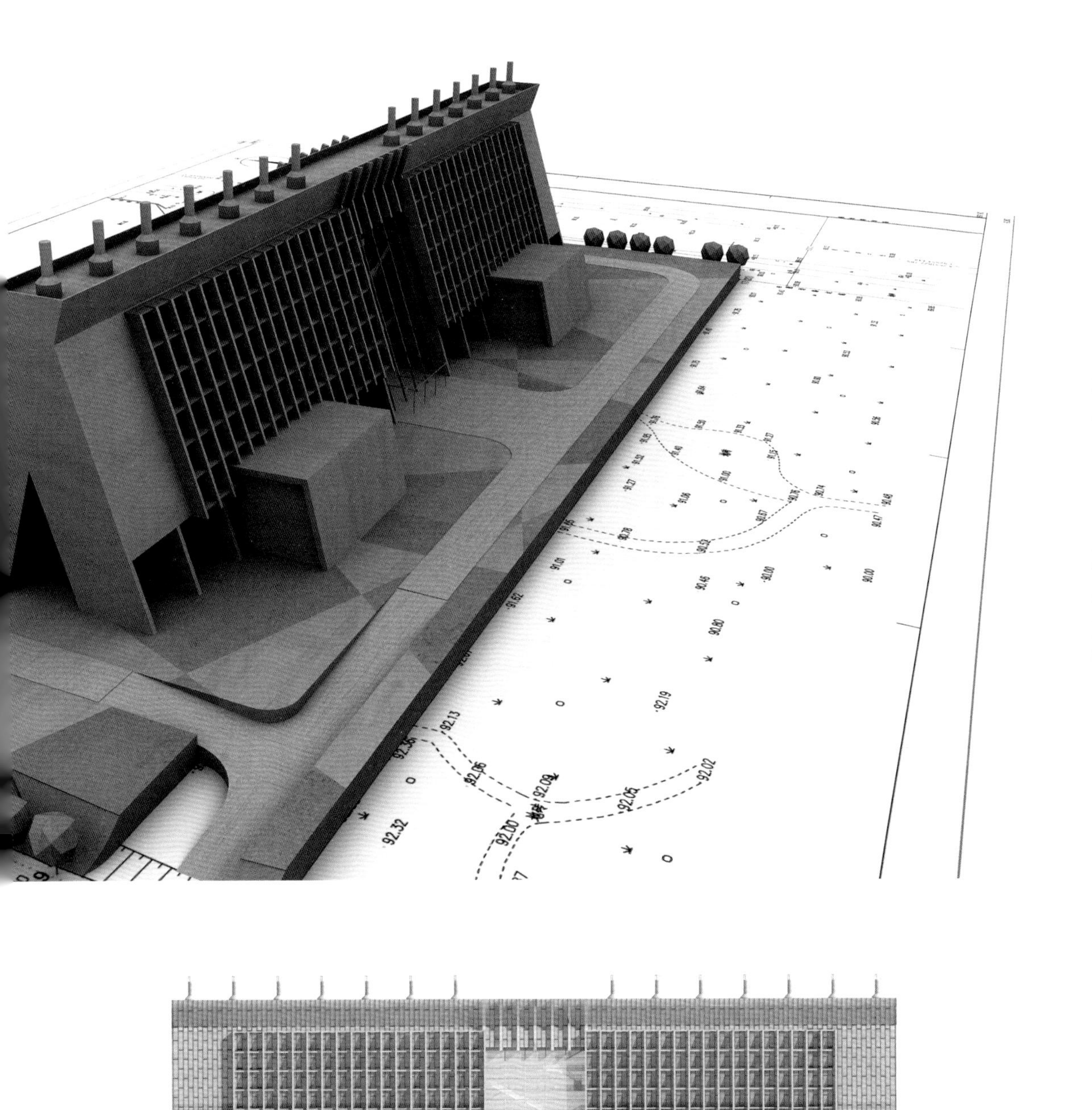

成都南站概念设计

Conceptual Design of Chengdu South Railway Station

主题：“无数春笋满林生”

Theme: “The Birth of Numerous Bamboo Shoots All Over the Forest”

成都南站位于成都市天府大道，在既有成都南站原地重建。成都南站的环境吵杂混乱，尤其是天府大道高架桥在车站站台上方跨过，对车站候车大厅造成不利影响。

本方案借用竹子作为造型主题，并运用现代的设计手法加以表达。竹子与成都的地域文化和人们的日常生活息息相关。久居成都的唐代大诗人杜甫著有“无数春笋满林生”的诗句，正是本方案欲表达的竹笋破土而出、竹林迎风摇曳的意境。

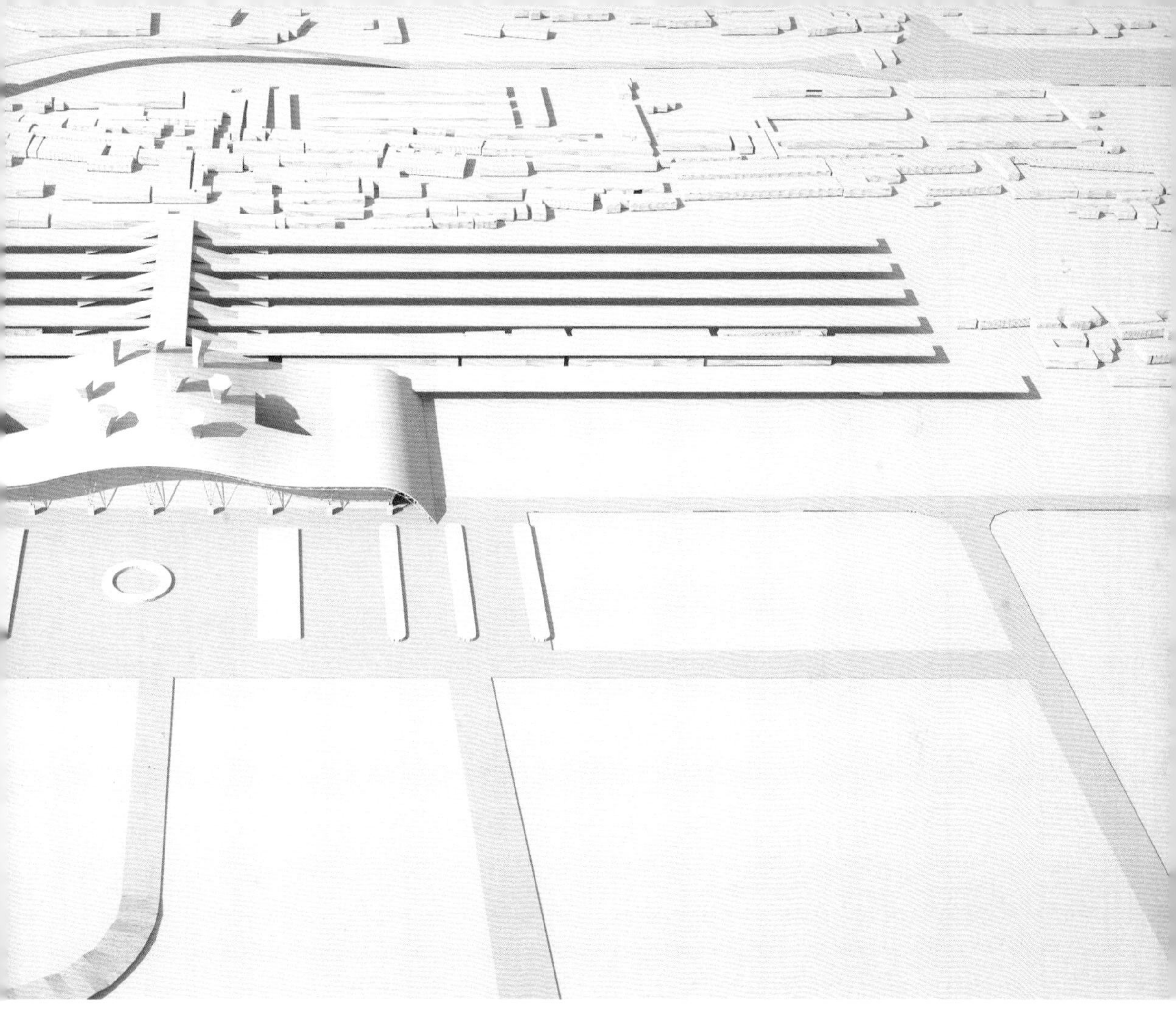

为了减轻高架桥对候车大厅的影响，建筑两侧屋面坡向地面，避开与高架桥的冲突。同时特别注意了第五立面——屋顶的设计效果。

The Conceptual Design of Chengdu South Train Station is a renovation of the existing Chengdu South Railway Station on Tianfu Avenue. The environment around Chengdu South Railway Station is noisy and chaotic, putting the station in advantage; especially the overpass bridge lay directly above the platform.

The project uses bamboos as its modeling reference, and creates a expression with contemporary designing techniques. The local culture of Chengdu, as well as the people's everyday life, cannot be separated with bamboos. Du Fu, a great poet of the Tang Dynasty who resided in Chengdu for extensive period

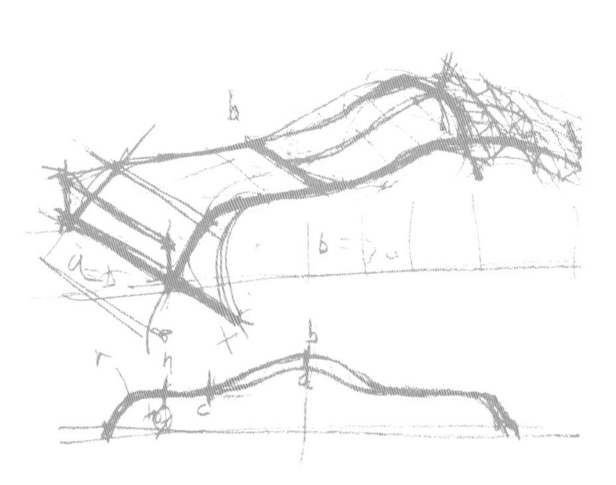

of times, wrote the framed phrase -"the birth of numerous bamboo shoots all over the forest". The project is about to present the picture described in the phrase, where a bamboo forest growing out of the soil and swinging along with the wind.

In order to reduce the impact by the overpass bridge to the platform, both sides of the architecture slope towards the ground. The

rooftop, as the fifth dimensional surface, received extra attention simultaneously.

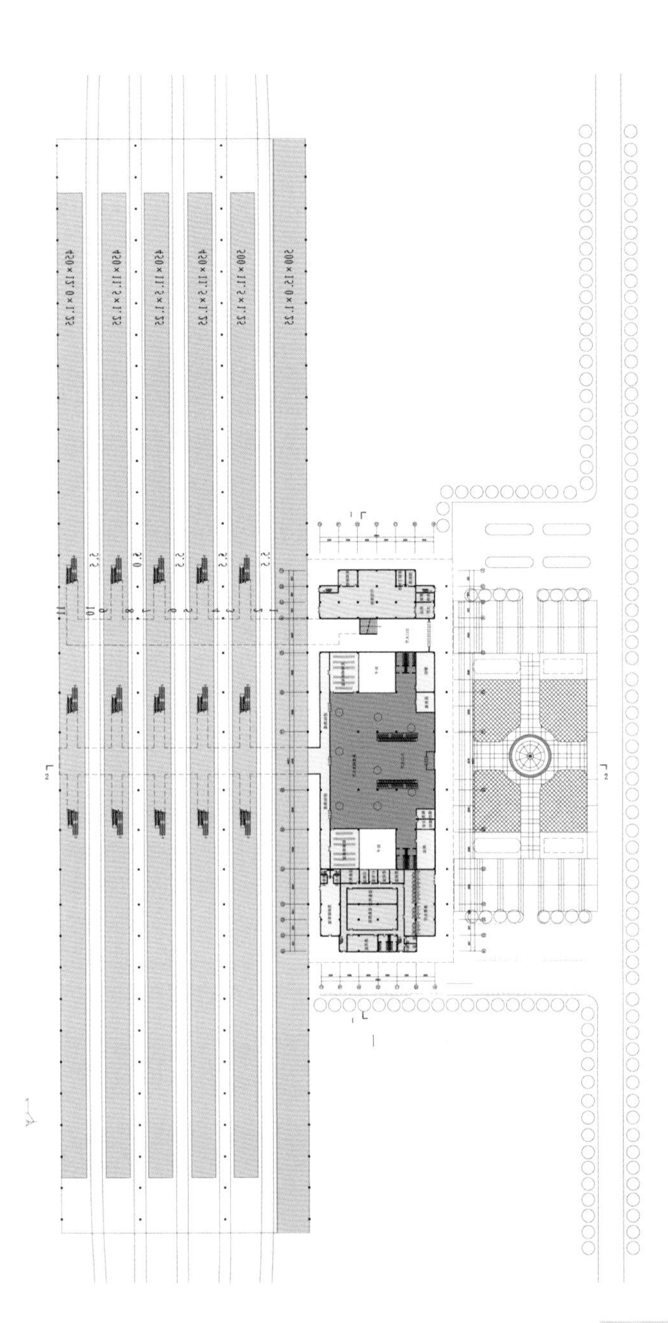

1

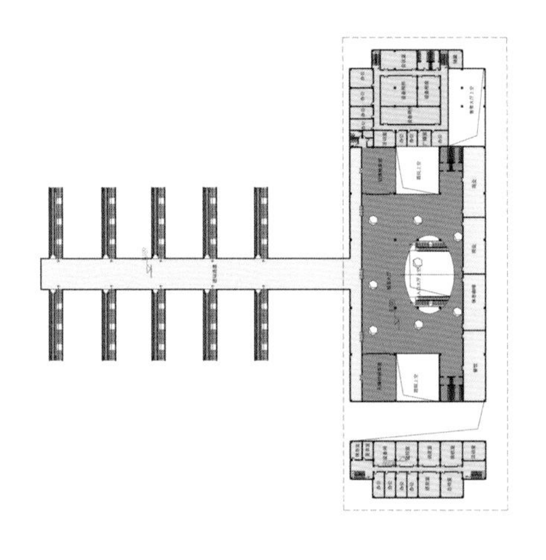

2

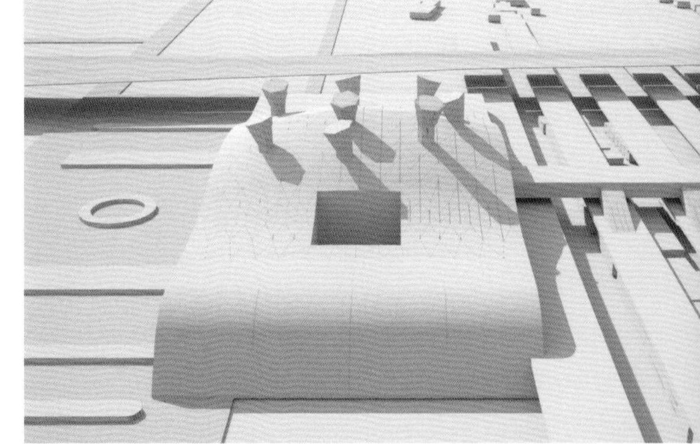

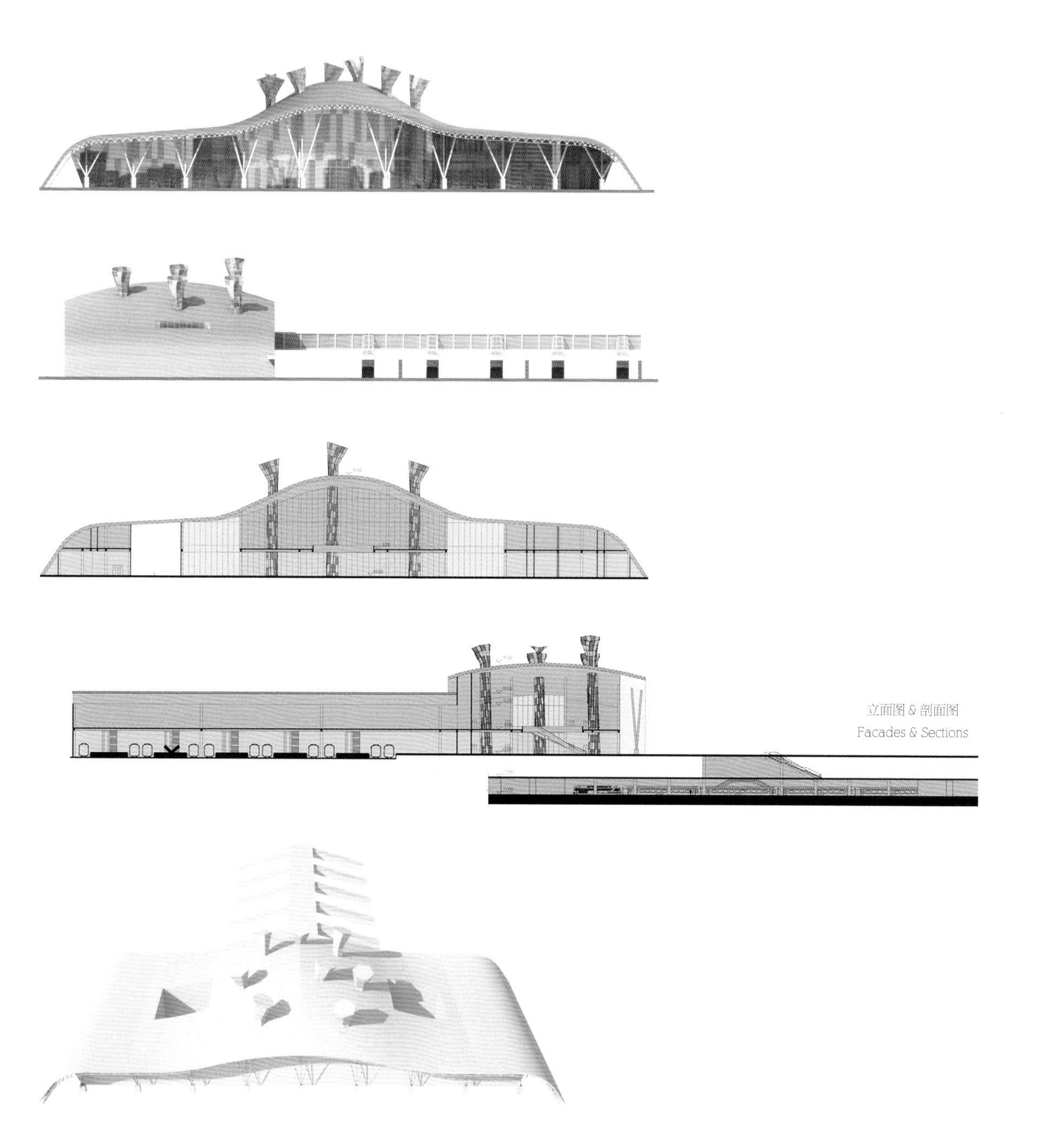

立面图 & 剖面图
Facades & Sections

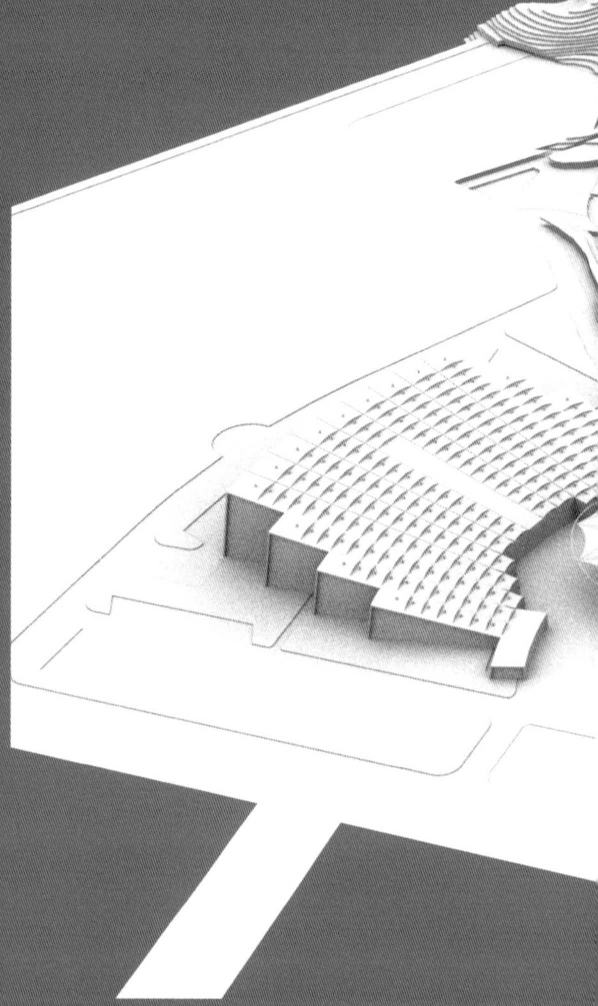

东兴国际展销中心

Dongxing International Exhibition Center

主题：会呼吸的屋顶

Theme: A Breathing Rooftop

东兴市位于广西西南，与越南仅一河之隔，是我国与东盟的重要贸易口岸。

东兴国际展销中心的设计秉承生态的理念。在总图布局上，考虑到当地气候与地形的特点，利用地段东北侧的水面和山体来改善建筑的微气候环境。在屋面设计上，借鉴“双层幕墙”，创造性提出“会呼吸的屋顶”的做法——在屋顶上覆盖一层“遮阳瓦”，在“遮阳瓦”尾部设有拔风塔，利用拔风效应带动空气流动，降低阳光的热辐射热。“会

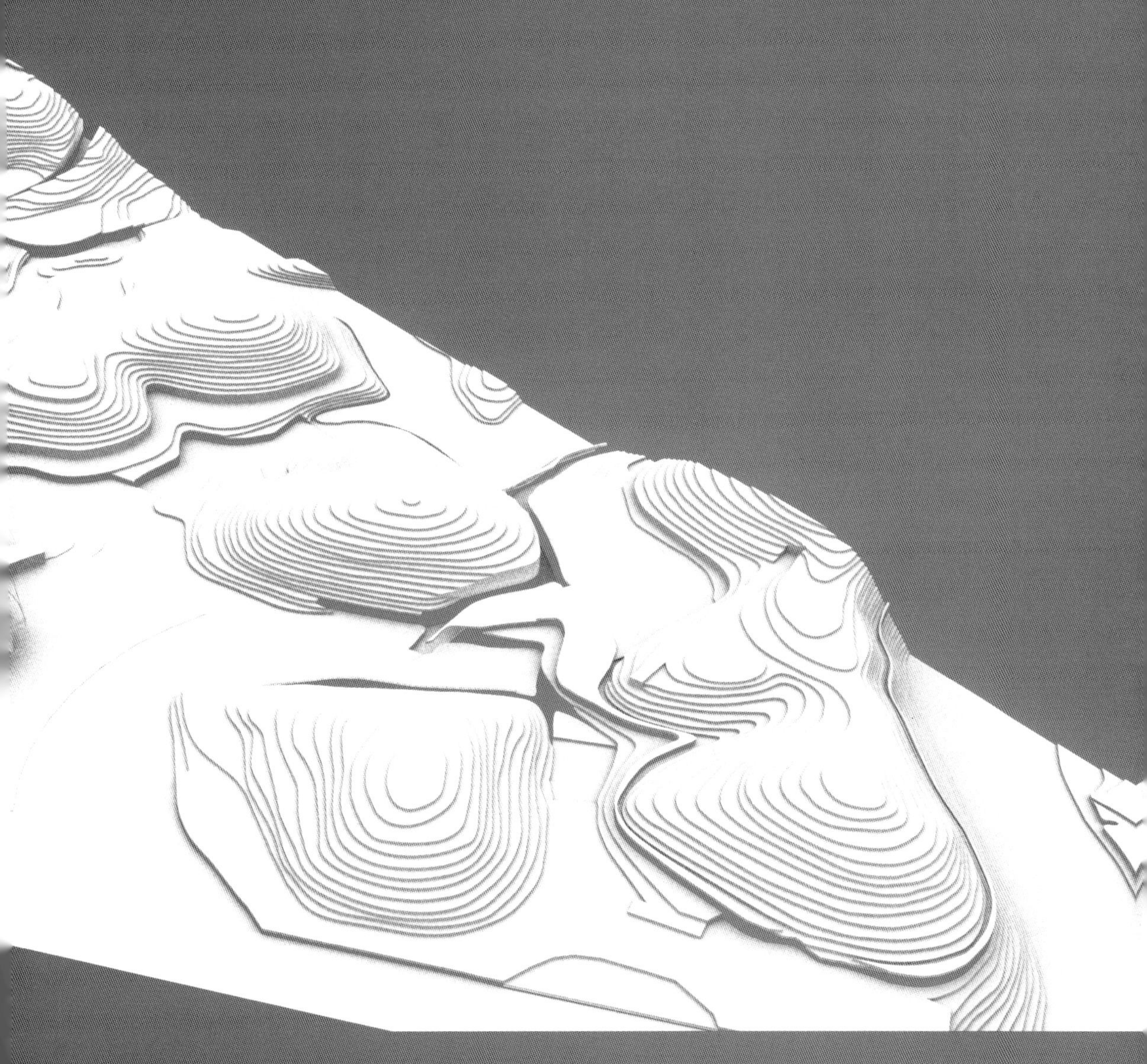

呼吸的屋顶”成为建筑造型的重要特征。

Dongxing City, located in the southwest of Guangxi and only a river apart from Vietnam, is an important commerce port between China and ASEAN nations. Dongxing International Exhibition Centre is designed adhering to ecological consciousness. Considering the characteristics of local climate and geography, water body and mountain body northeast of the area were used to improve the micro-climate in the blueprint layout In the surface design of the architecture, an innovative "breathing rooftop" was create referencing the "double layered curtain wall" design. A layer of "sun shading tiles" covers the rooftop. The "sun shading tiles" lower the radiation heat from direct sunlight by flowing the air with the

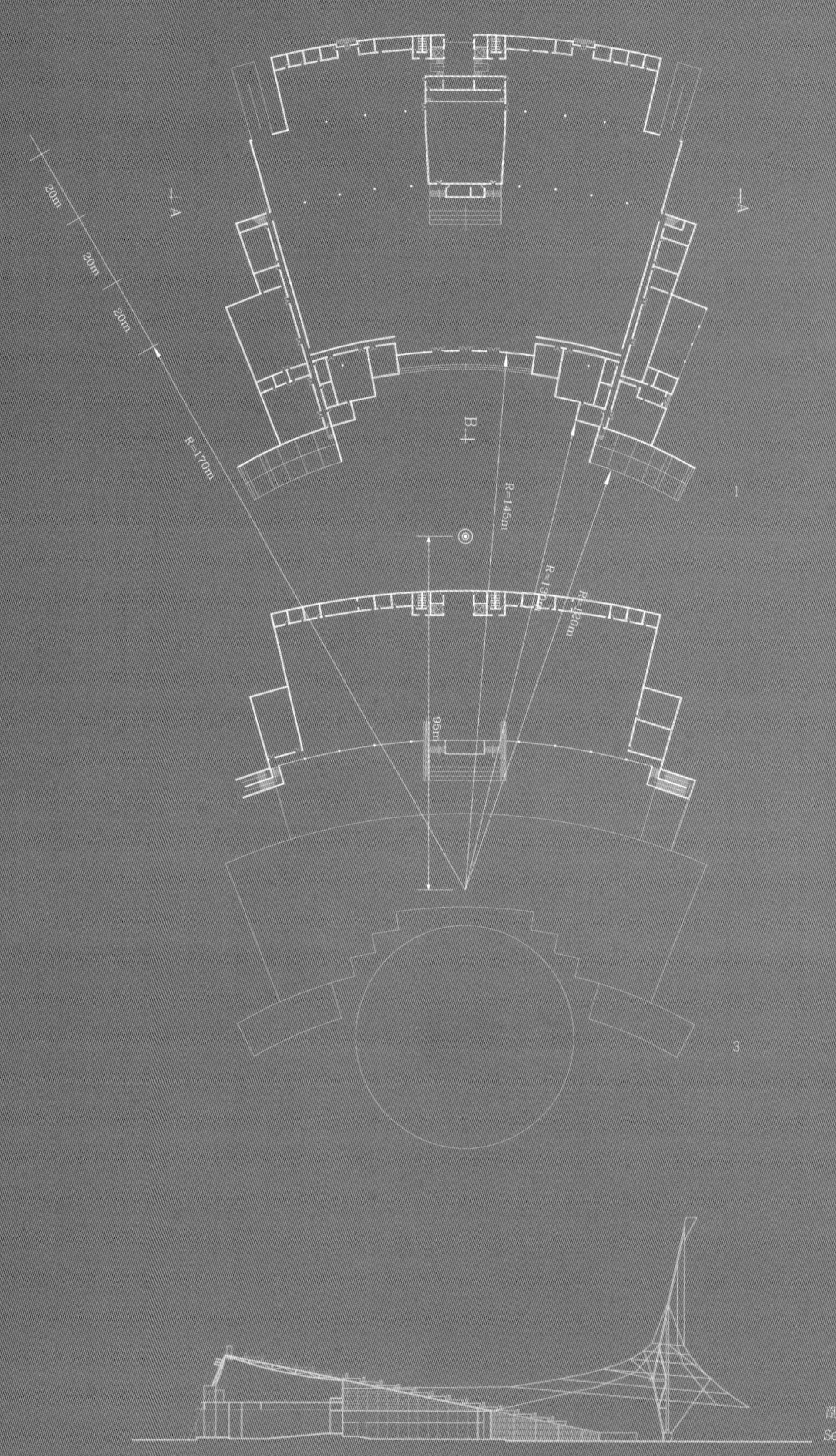

剖面图
Section

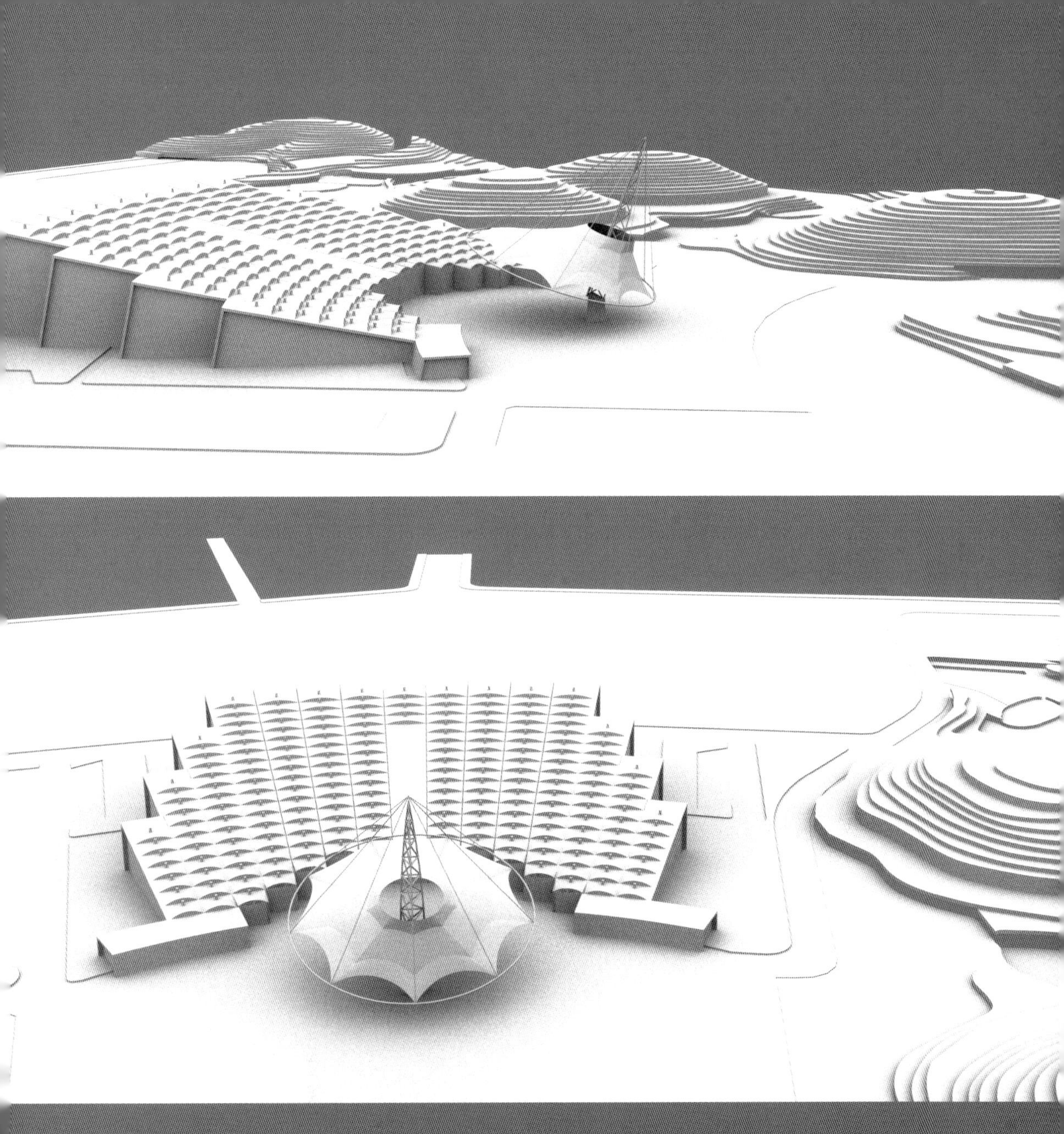

vacuuming wind tower on their tail. "A breathing rooftop became an essential traits of architectural model.

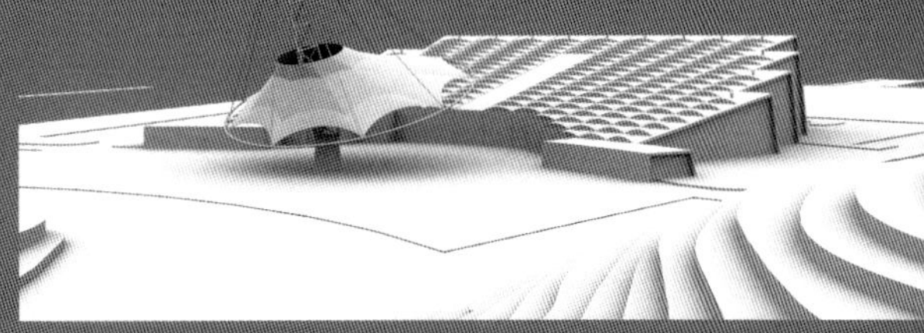

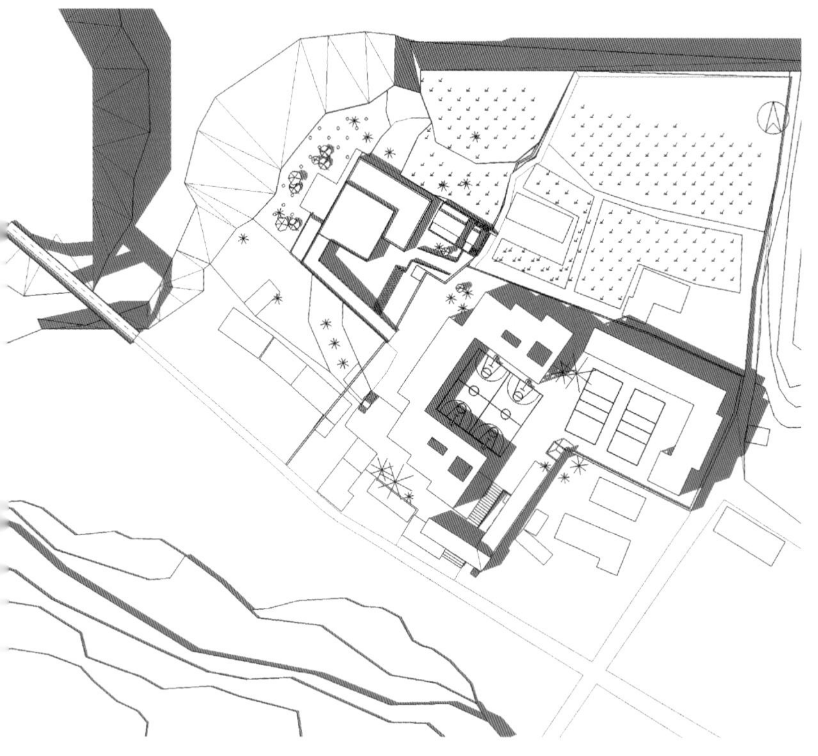
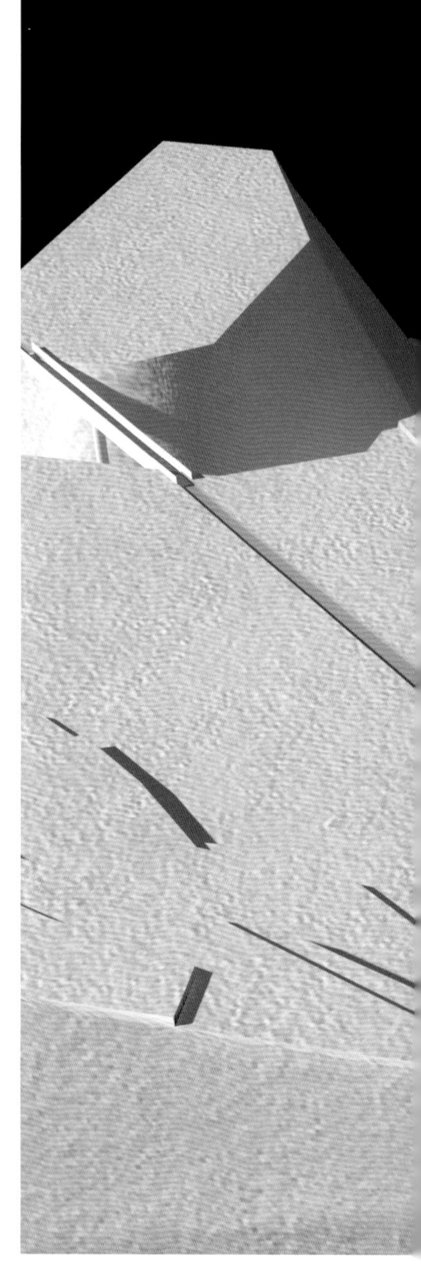

文县横丹学校
Wen County's Hengdan School

主题：建构本土化

Theme: Tectonics Localized

横丹学校是 2008 年汶川大地震后，北京市电力总公司捐赠的学校。横丹乡位于甘肃省秦巴山区深处，校园用地非常局促，仅仅 6700 平方米，并且用地高低不平，最大标高相差近 5 米。

用地严重不足是学校重建面临的主要问题。为此本方案首先将建筑沿地段周边布置，尽量留出中间活动场地；其次，将教学楼底层架空，走道加宽，为学生提供了一个遮风避雨的半室外活动场地；再有，将部分建筑屋顶设计成上人屋面，并通过坡道

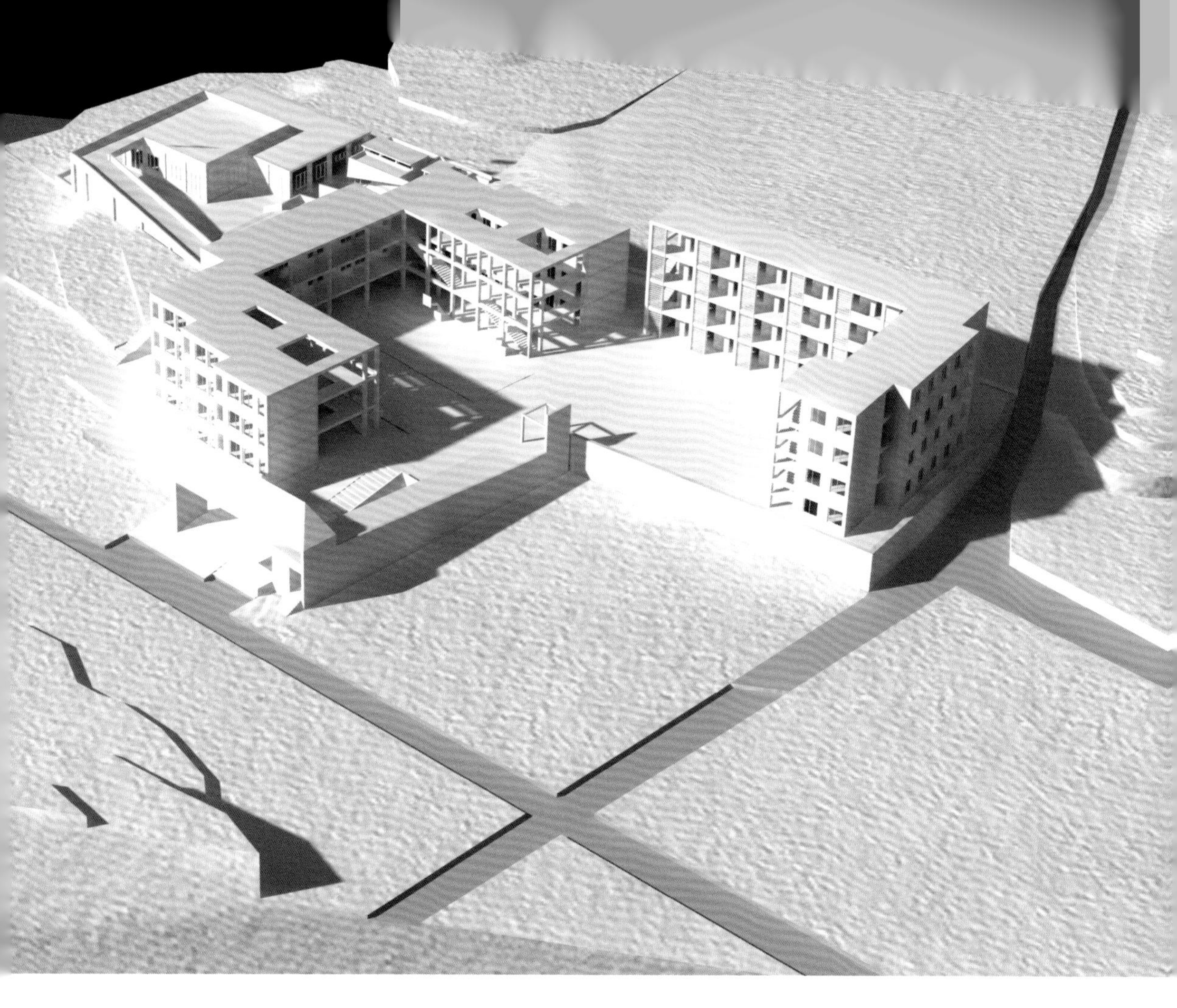

与地面场地相连。

地域化和生态化是建筑界的热门话题。而在地域偏远、发展落后的地区如何实现这个目标是一个值得思考的问题。由于缺乏活动场地，本方案采用了平屋顶的做法，因为对于学校，活动场地比漂亮的坡屋顶更重要；由于当地缺水，本方案放弃了冲水马桶加分散式排污的生态设想，而是采用了旱厕，但在卫生条件和粪便运输方面做了改善；由于自然资源缺乏，本方案放弃了黏土砖和木质格栅等时下流行的地域材料做法，而是采用了当地工业废渣制作的砌块。地域化和生态化要适应地方客观条件，而不是包装上看似地域或生态材料的外衣。

The Hengdan School is a donation made by the Beijing Electric Power Corp. following the earthquake in Wenchuan during year 2008. Hengdan Xiang is located deep in the Qinba Mountain District, Gansu Province. The school

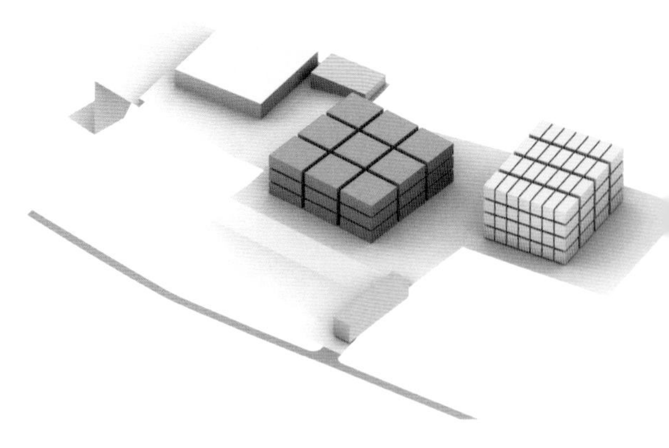

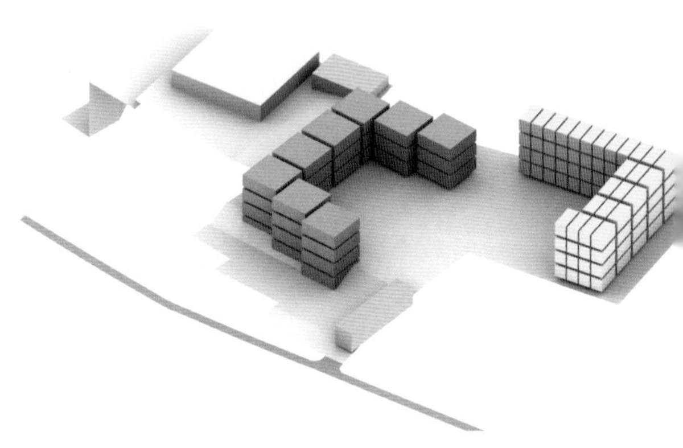

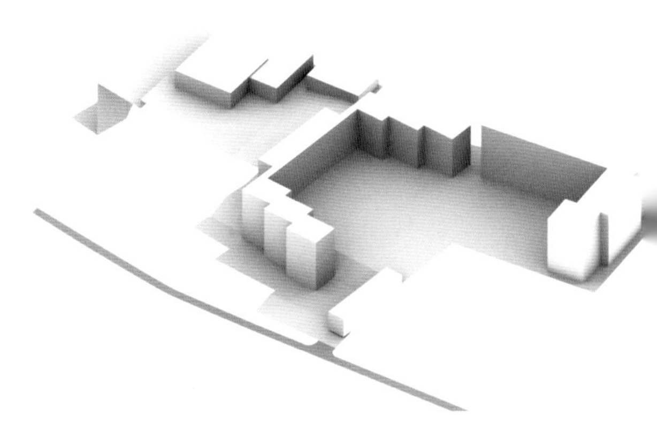

1. 地震前 Before the earthquake
2. 地震后 After the earthquake

is built in a cramped area of land, netting only 6,700 hundred square meters. The land is also not levelled with the maximum height difference of 5 meters.

The major problem the project encountered was the critical lack of land. The project first arranges uses for the edge of the land, focusing on leaving maximum space open in the middle; then the project provides students with

a weather resistant, semi-open space by widening the walkways and hollowing the ground floor of the lecture hall. Access to the rooftop is also granted via ramps from the ground level.

Localization and ecology is a hot topic in the architecture industry; but how to carry out the above in remote and underdeveloped area is the valuable question. It is because of the lack of playing fields, the school took priority of a flat roof over a magnificent sloped roof to achieve more accessible spaces. Water-flushing toilets and distributed sewage are given up for the lack of water in the location. Dry toilets were adopted with improvements on sterilization and transportation of feces. The project utilizes local factory waste as building materials, instead of the popular bricks and woods, because of the lack of natural resources.

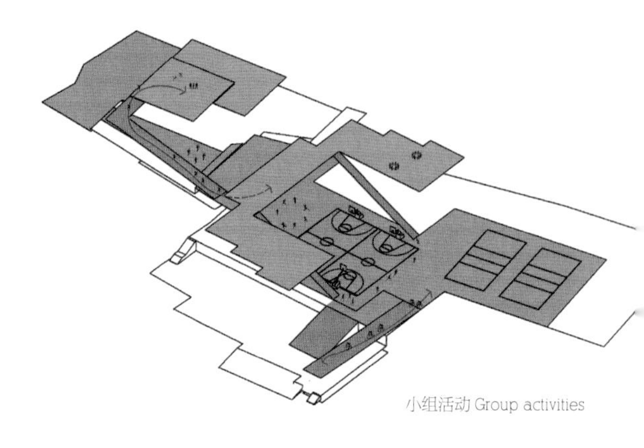

小组活动 Group activities

1

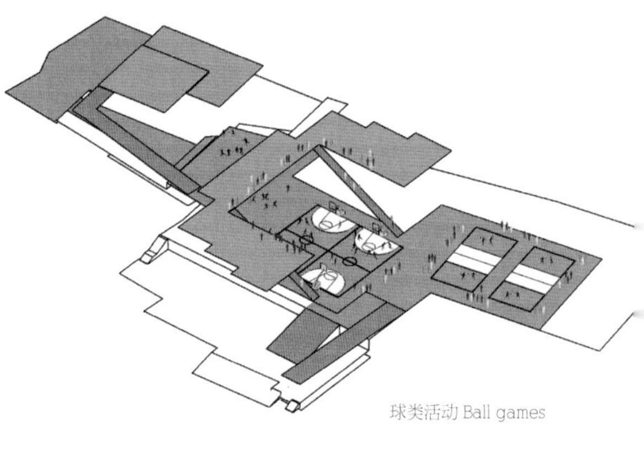

球类活动 Ball games

2

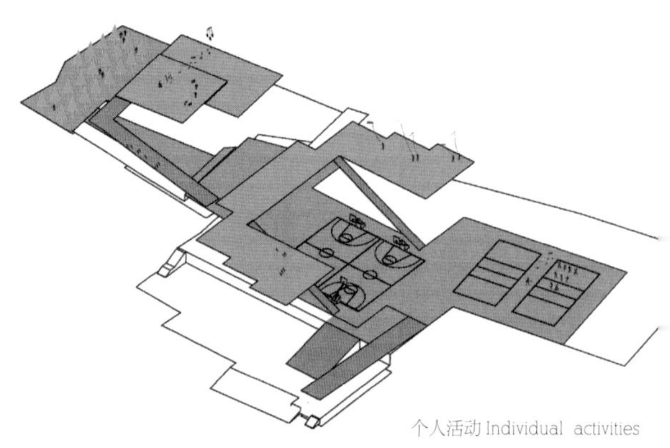

个人活动 Individual activities

3

1. 援建者和学生 Donator with the students
2. 临时校舍 Temporary classroom
3. 露天上课 Having class outdoor

Localization and ecology is not about wrapping the area with pseudo local or ecological material, but to adapt to local conditions.

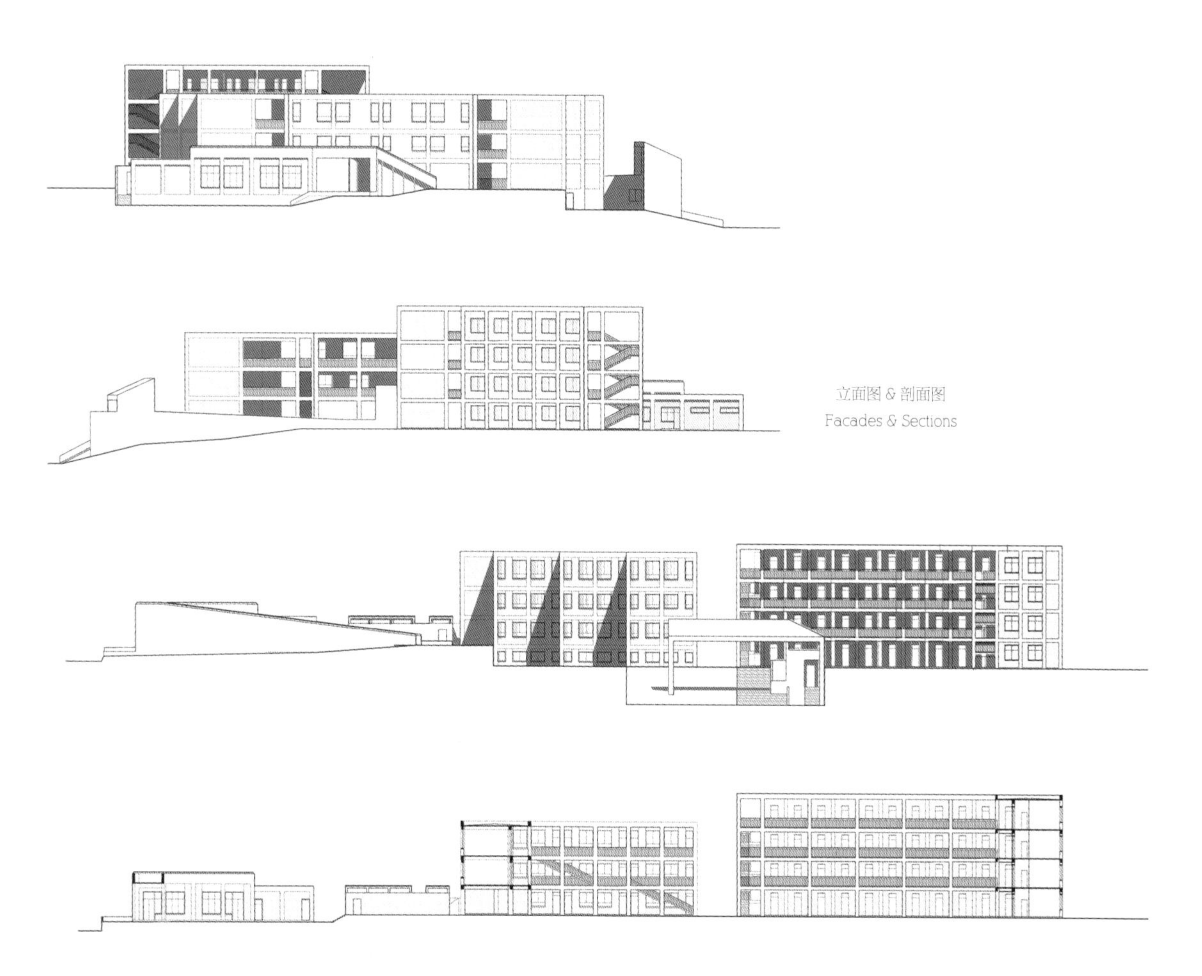

立面图 & 剖面图
Facades & Sections

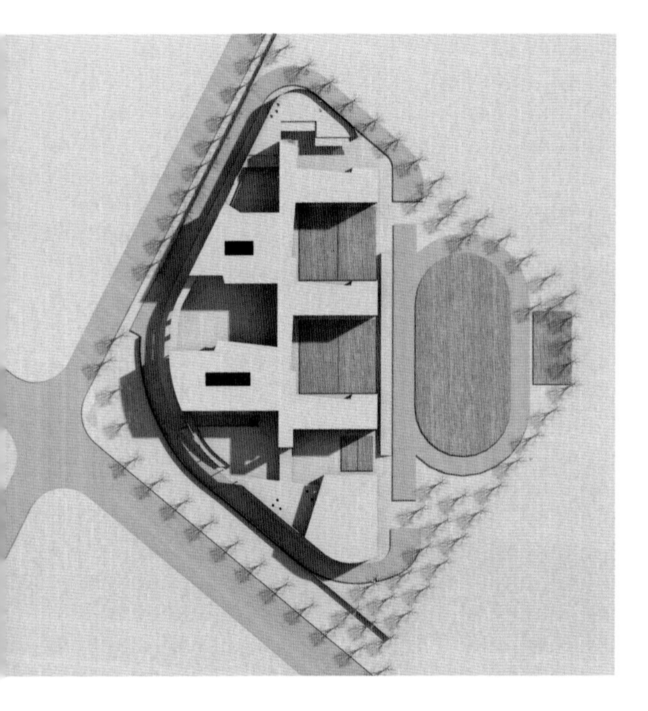

古亭山小学

Gutingshan Primary School

主题：正南正北

Theme: All Face to South and North

古亭山小学的整体空间结构可划分为建筑群部分和绿化活动场地部分。建筑群相对集中在用地西侧，由一条主干连廊贯穿南北，将各栋建筑连接起来。大型运动场场地相对集中在用地东侧，小型绿地和活动场地穿插布置在建筑之间。

这样布局产生的结果是，运动场地按正南正北方向布置，完全满足设计规范对体育场地的要求；建筑群亦按正南正北方向布置，具有良好的朝向和充足的采光。主干连廊完全开敞，教学楼单侧外廊

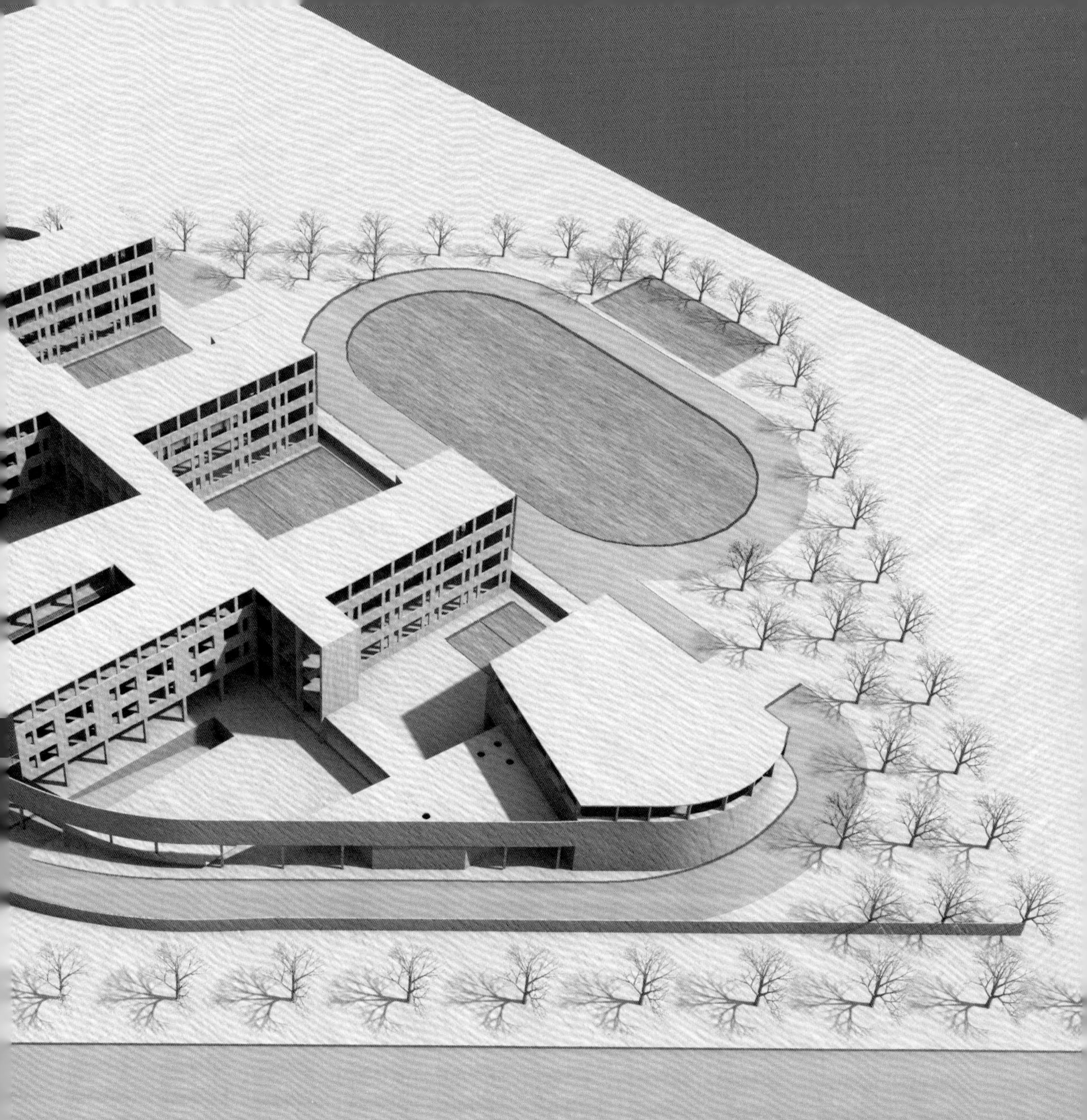

敞开，办公楼和综合楼的底层架空。通过这些措施，到达良好的通风效果和卫生要求。为了避免外部噪声的干扰，本方案结合建筑外墙和校园围墙，设置了两道大型弧形隔音墙。

The overall space structure of Gutingshan Primary School can be divided into the architecture zone and the green zone. The architectures are relatively concentrated in the west side of campus. A main corridor links the buildings from south to north. Larger fields and gyms are built on the east side of campus, where playgrounds and micro green zones are arranged in between.

As a result, facilities of athletics face to the south and north, fulfilling the requirements by the design specifications on the sports fields . Other architectures also face to the south and

开敞廊道系统
Open corridor system

隔声墙
Soundinsulating walls

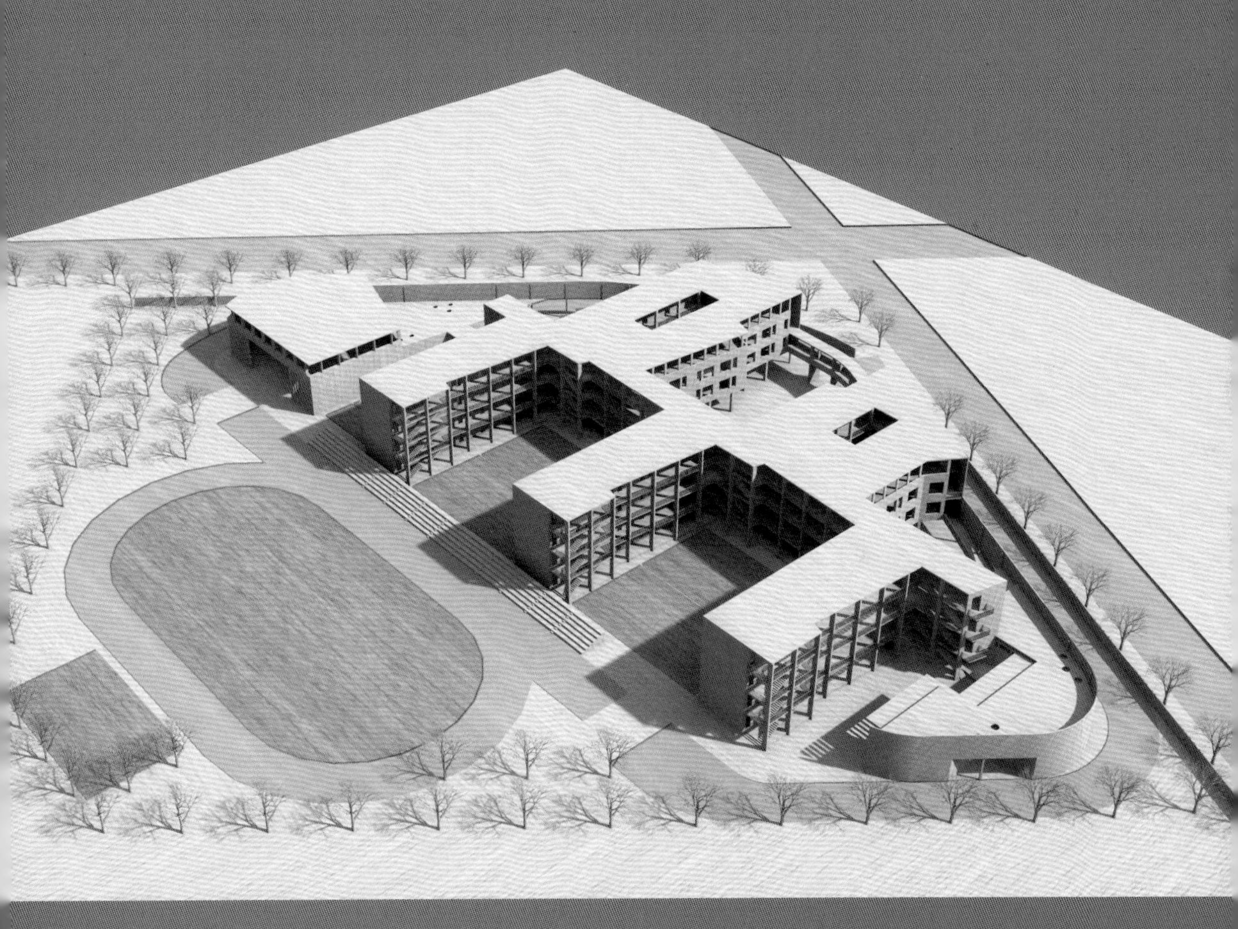

north, achieving great orientation and sunlight. The main corridor is completely open while walkways of the lecture halls are open from the side. The ground floors of office buildings and multifunctional buildings are hollowed. Great air ventilation and health condition are achieved by the above measurements. In order to eliminate excessive noise, the project merges the exterior walls and the campus fences, setting up two sound insulating walls.

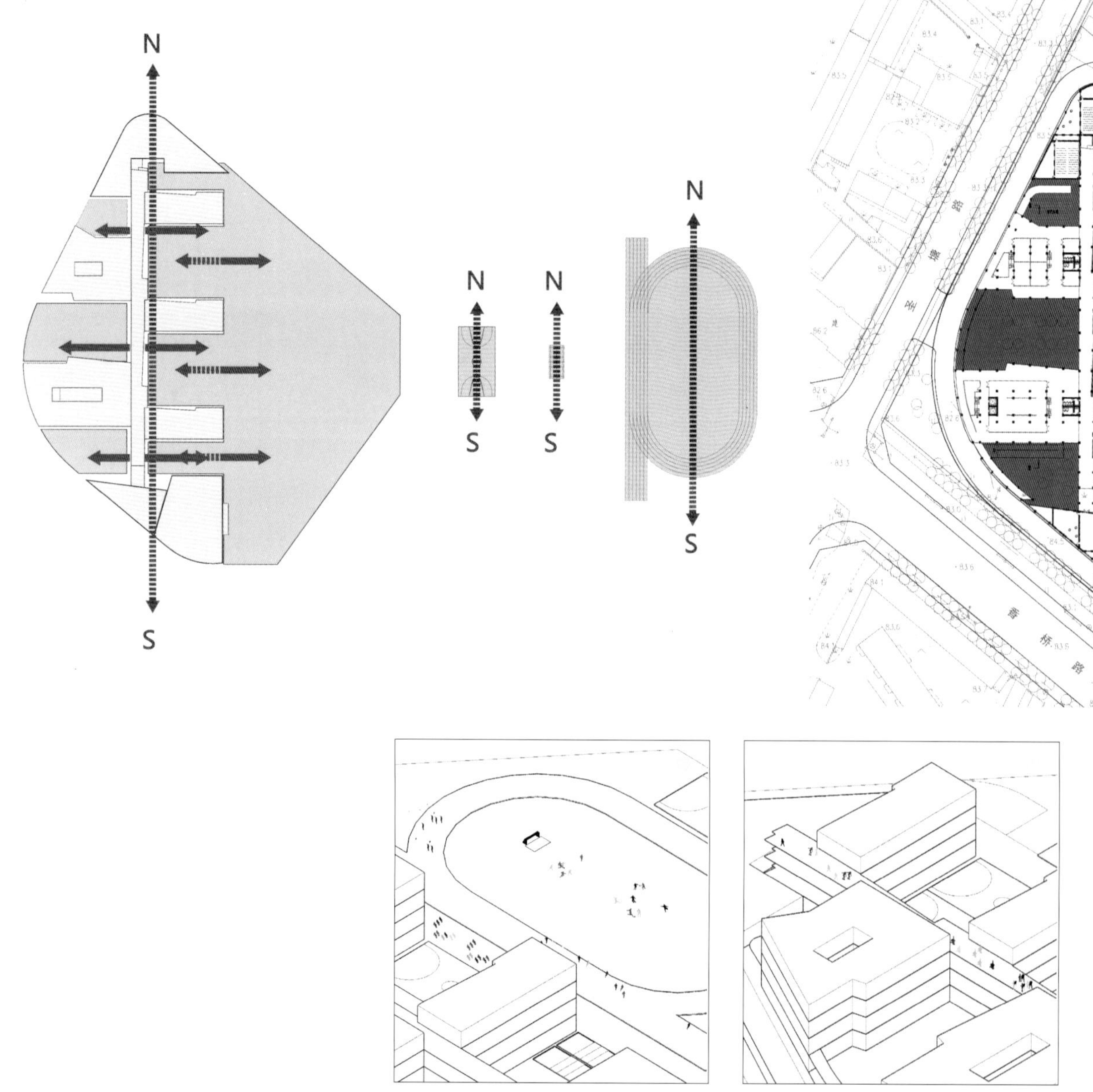
N
S
N
S
N
S
N
S

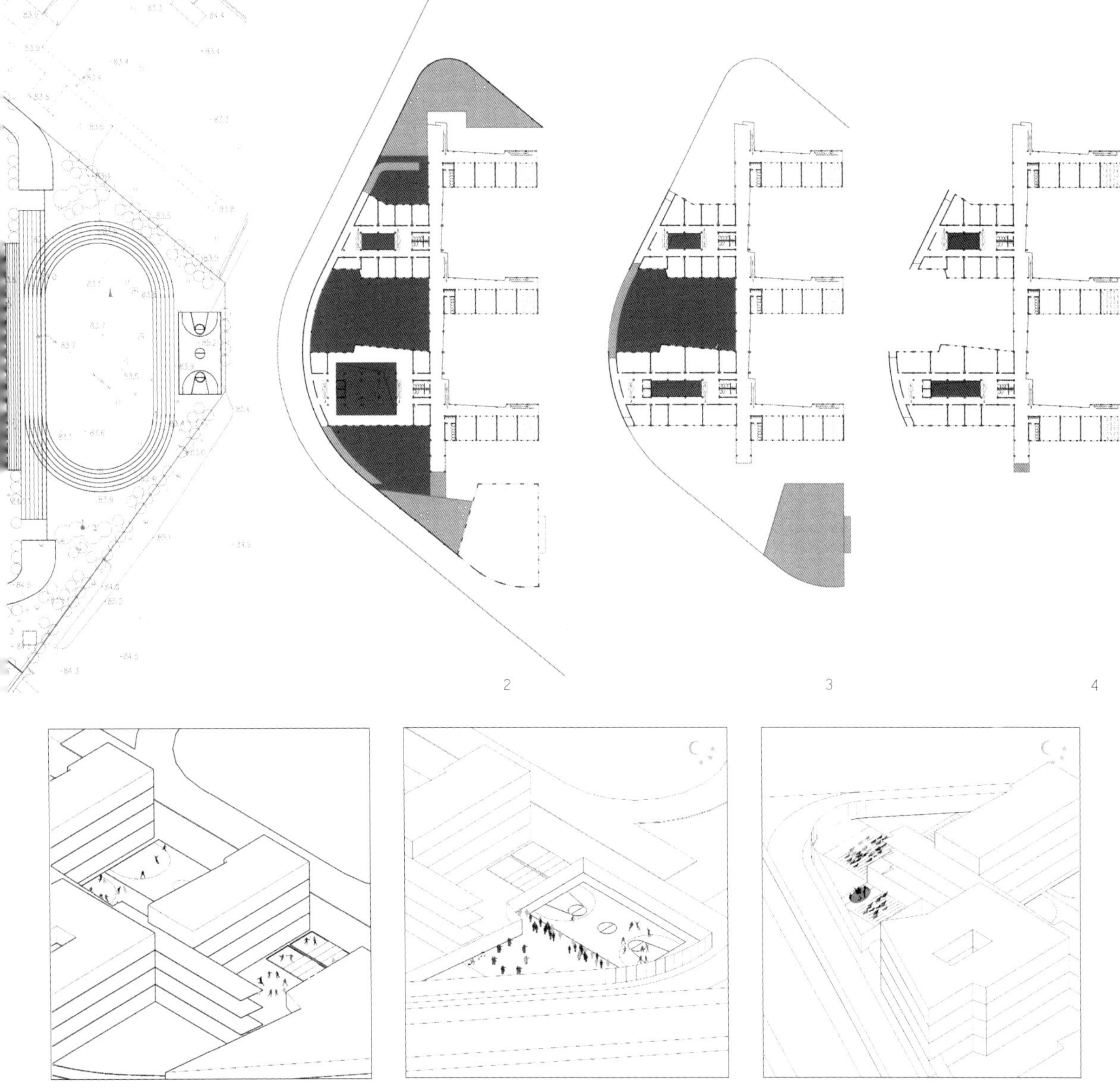
2
3
4

曦湾别墅

Xiwan Villa

主题：圆图腾

Theme: Round Totem

曦湾别墅位于内蒙古鄂尔多斯市，其总体布局吸纳了蒙元文化中“圆图腾”的理念。蒙古族漂泊的生活中，到处充满了“圆”。蒙古包、敖包是圆的，人们相聚在一起也围成圆。“库仑”一词在蒙语中表示圆圈、牲口圈舍、围墙，而“赞扬”、“同意”等词是由“库仑”演变而来。“圆”代表一种自满自足的美，一种永恒的美。本方案将别墅群纳入圆形组团，不仅仅体现了地域文化，也使建筑在保持正南朝向的同时，与斜向道路具有良好的空间关系。

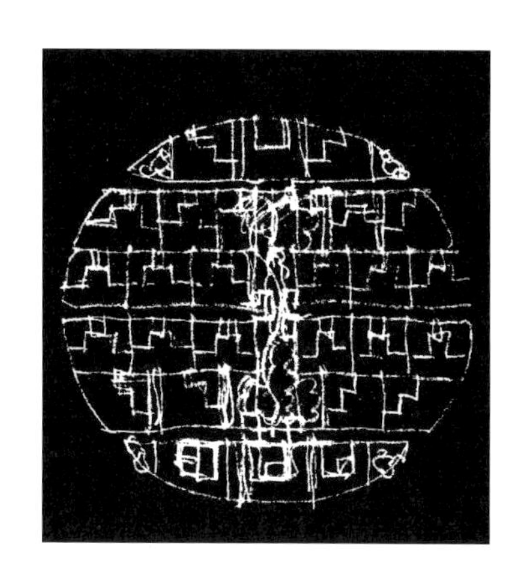

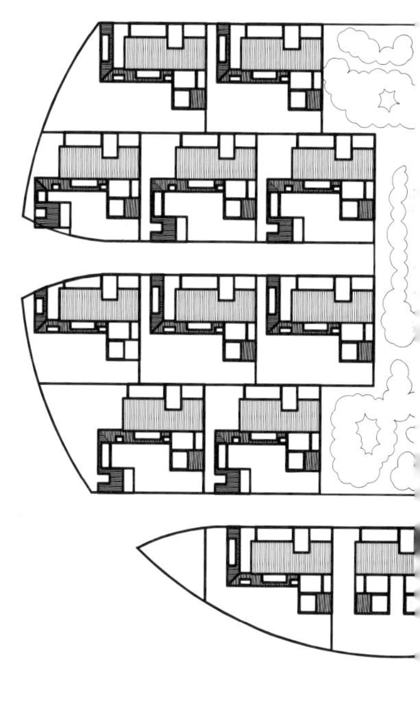

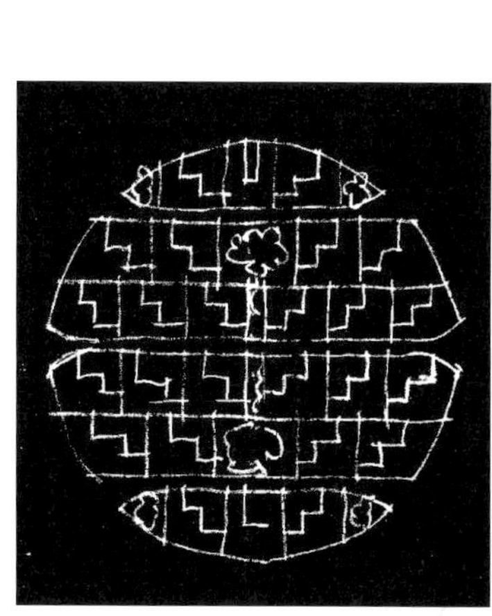

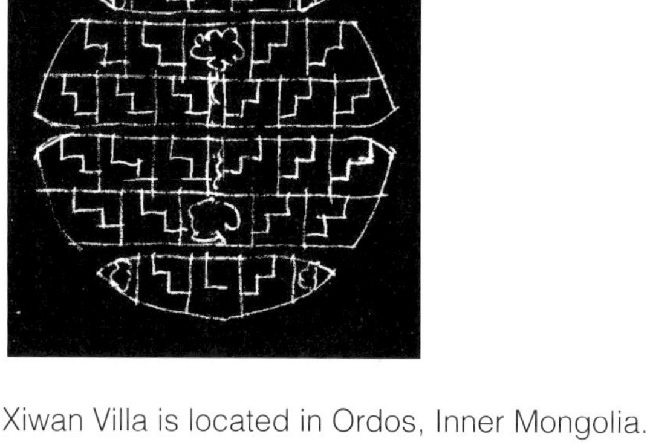

Xiwan Villa is located in Ordos, Inner Mongolia. The overall layout takes in the concept, "Circle Totem", from Mongolian culture. The drifting lives of Mongolians are full of "circles". Yurts and aobao are round-shaped, the people also group in a circle when gathering. The word "Ku Lun" carries the meaning of circle, as well as round shaped huts for livestock and surrounding walls; the words "praise" and "agree" are both derivatives of the word "Ku Lun". "Ku Lun" represents the eternal beauty of complacency. This project arranges separate villas into a circle shaped order while facing south; it does not only reflects the local culture, but also maintain good space between each villa adjacent to each other.

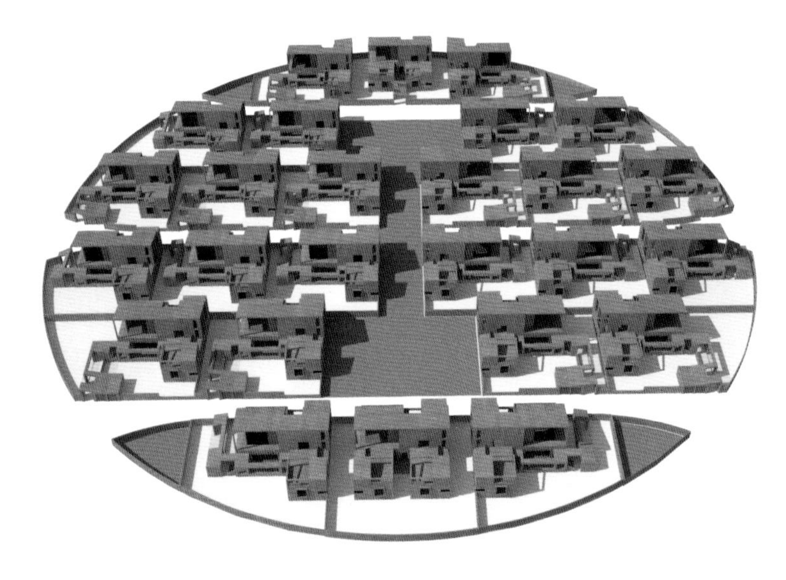

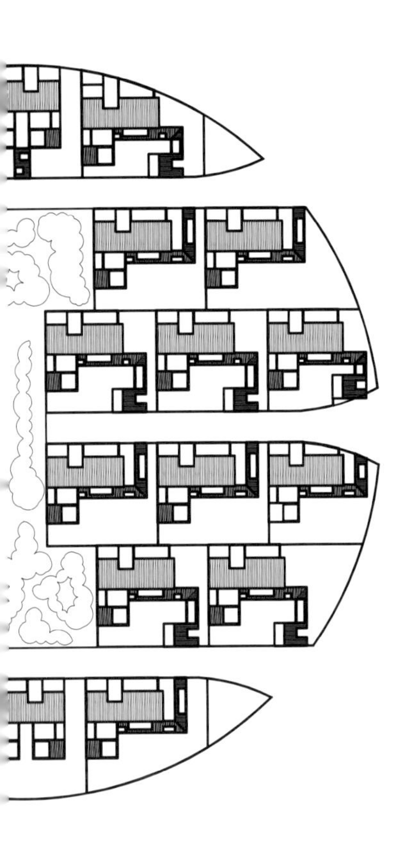

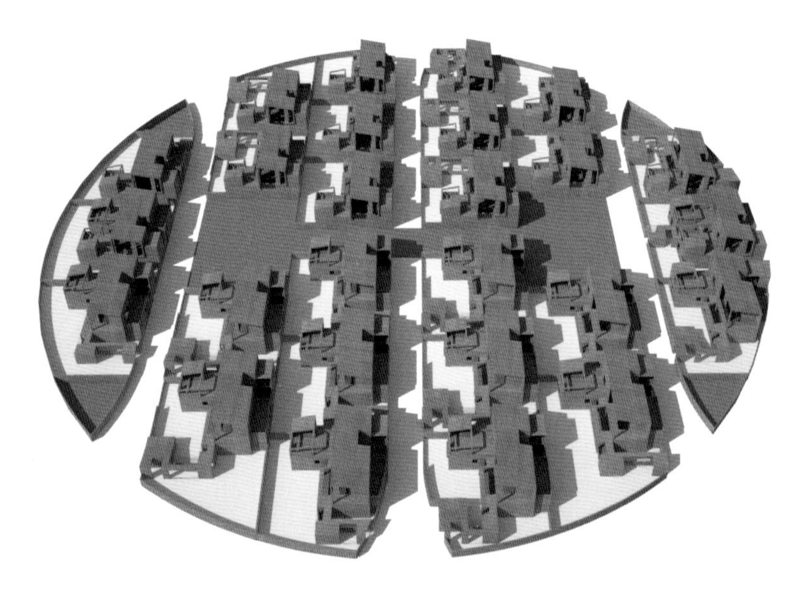

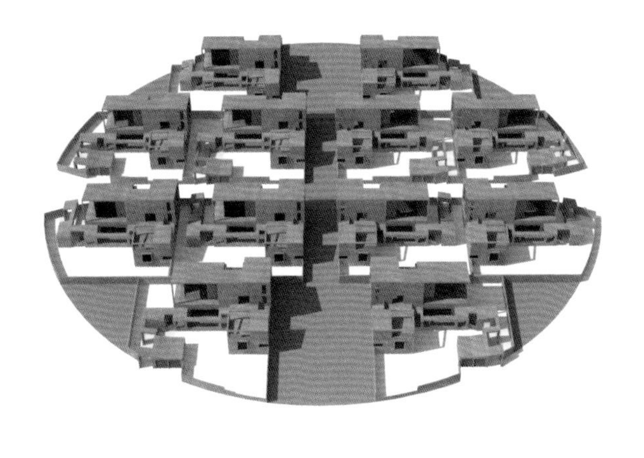

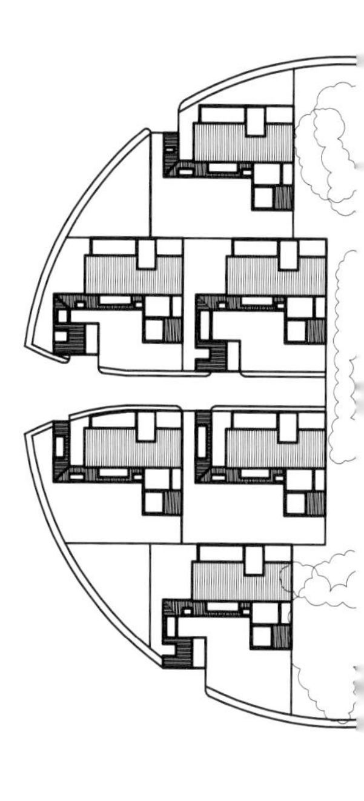

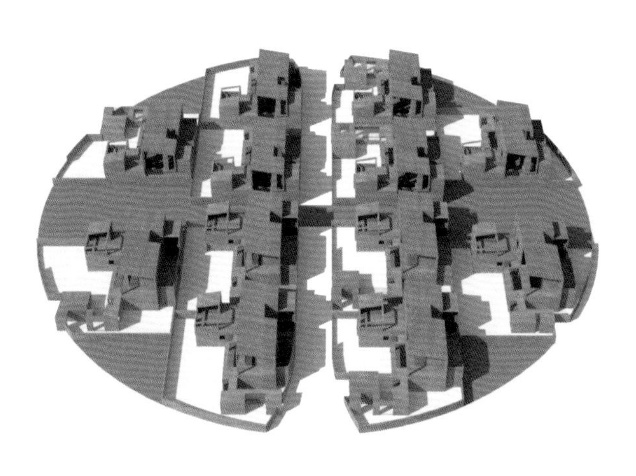

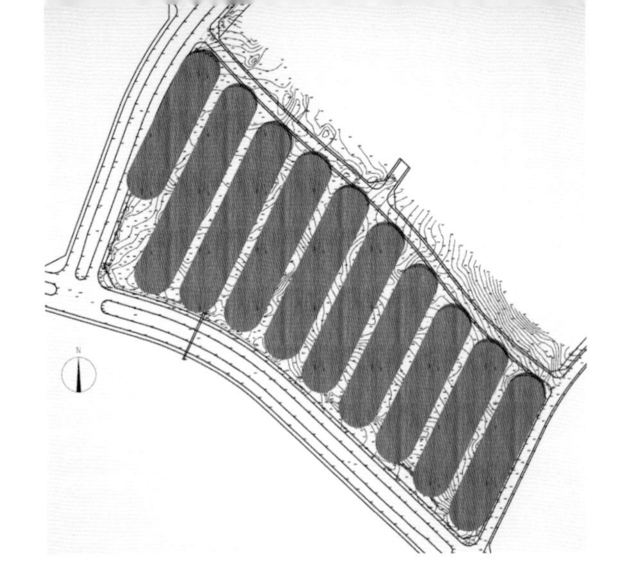
N

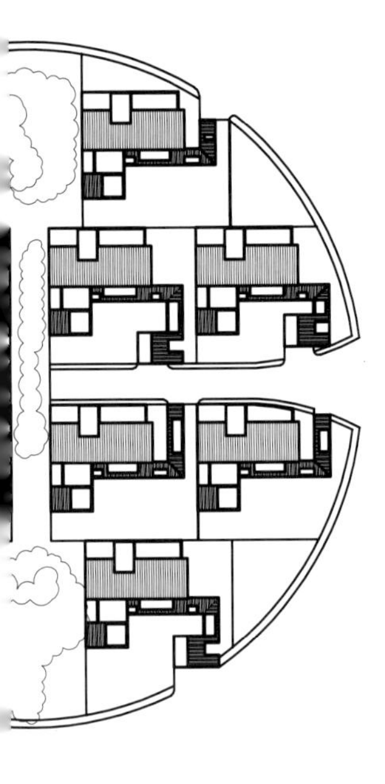

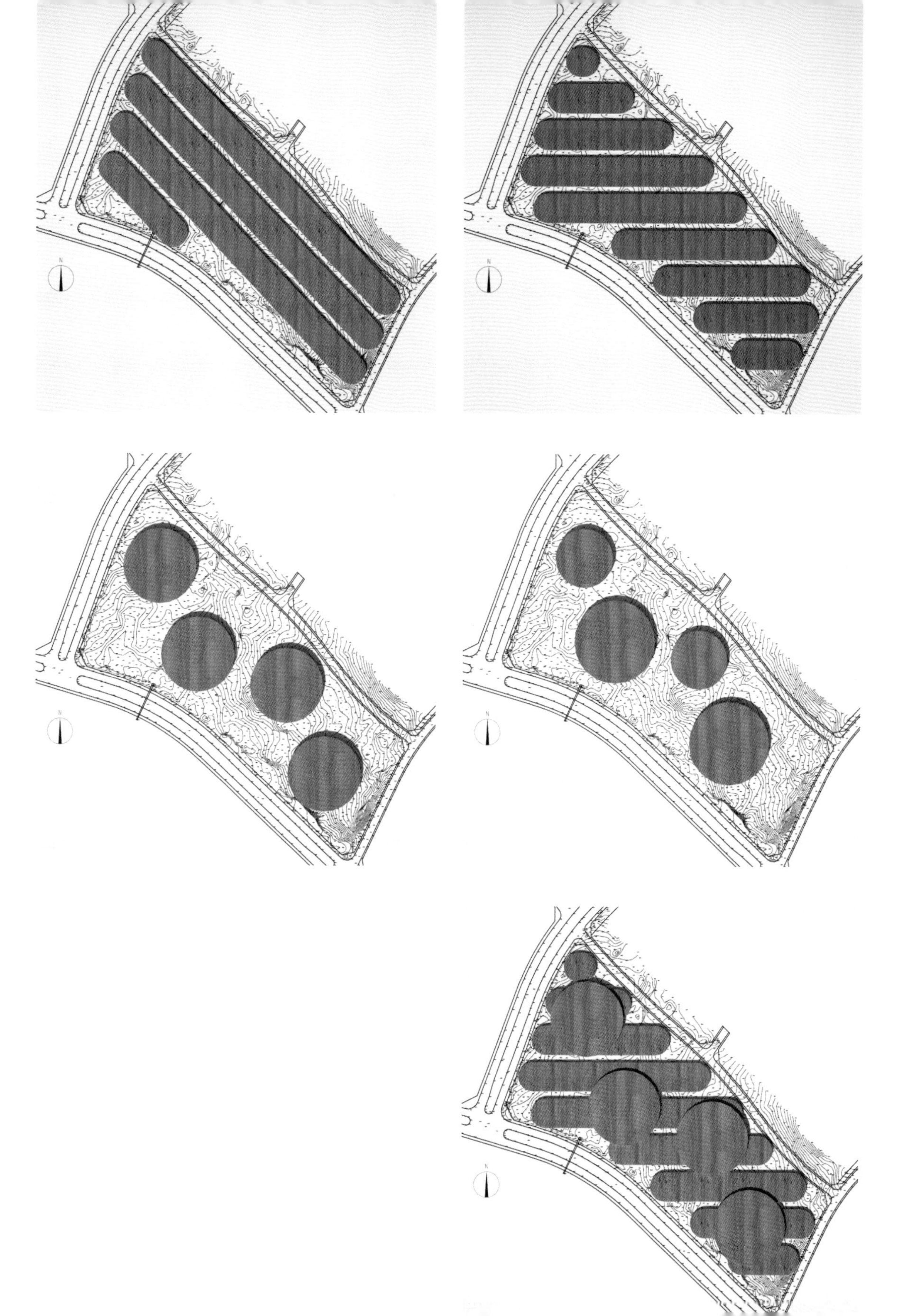

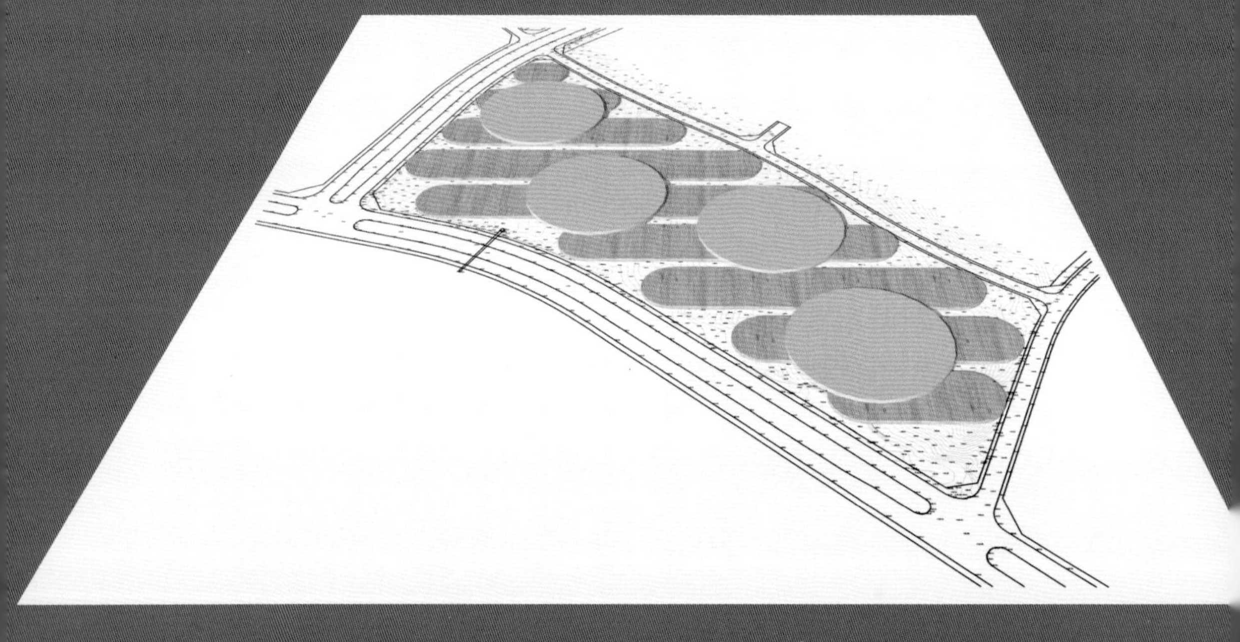

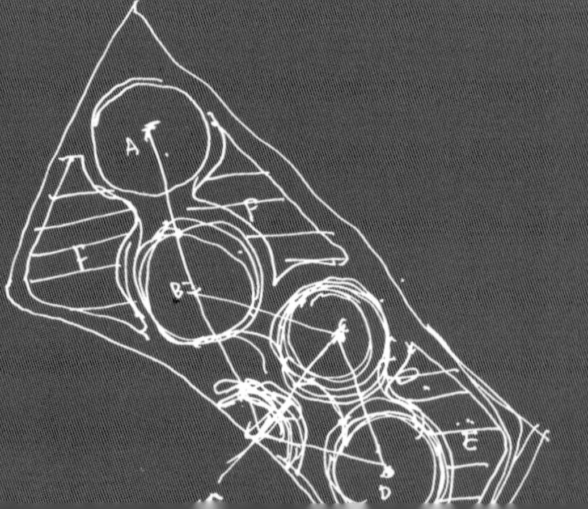
A
B
E
D

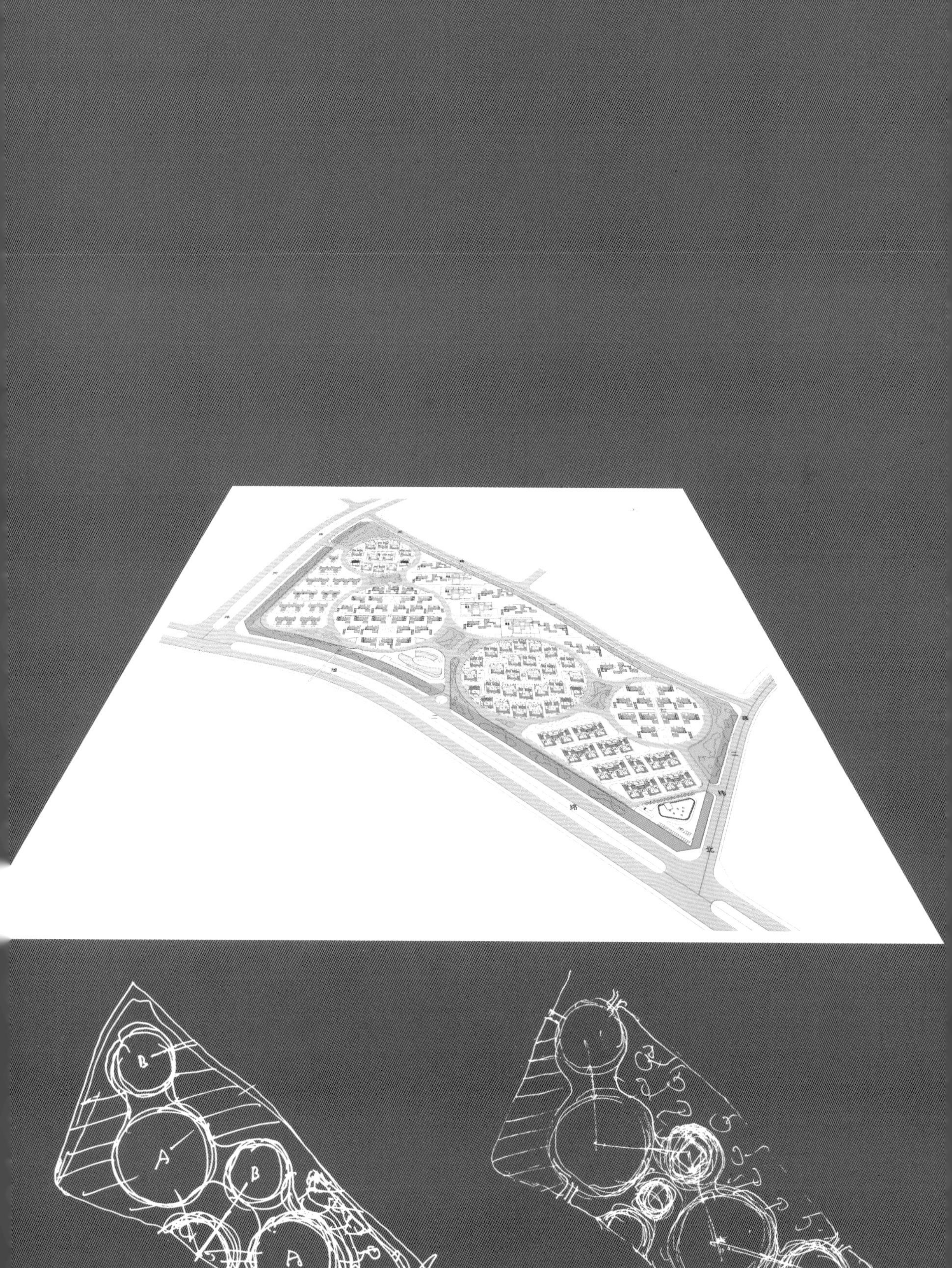
B
A
B
A

本土地景 LOCAL LANDMARK
文脉更新 CONTEXT RECONSTRUCTED
本土建构 LOCAL TECTONICS
城市空间 URBAN SPACE
本土再造 LOCAL UPCYCLING

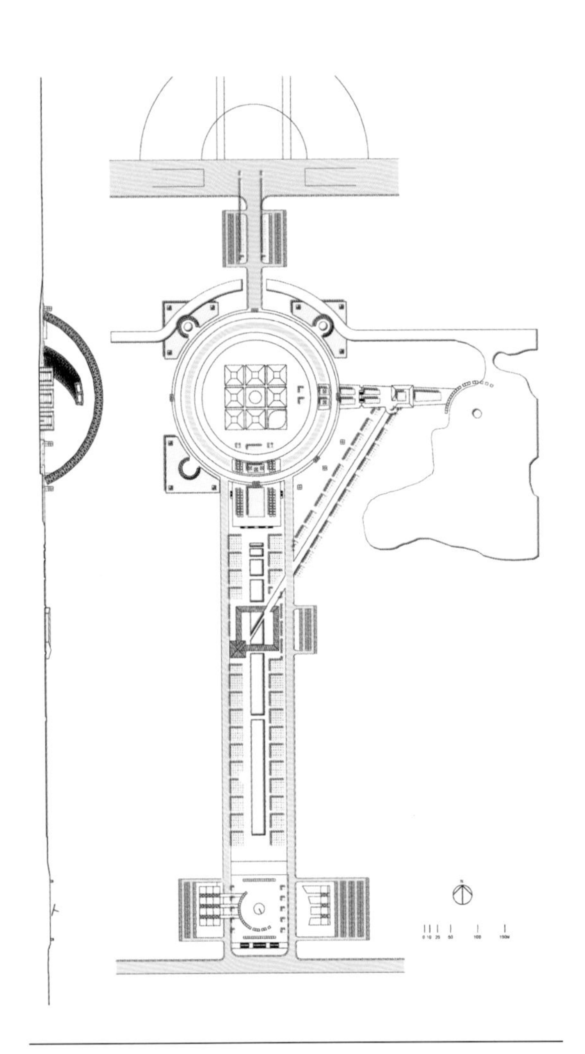

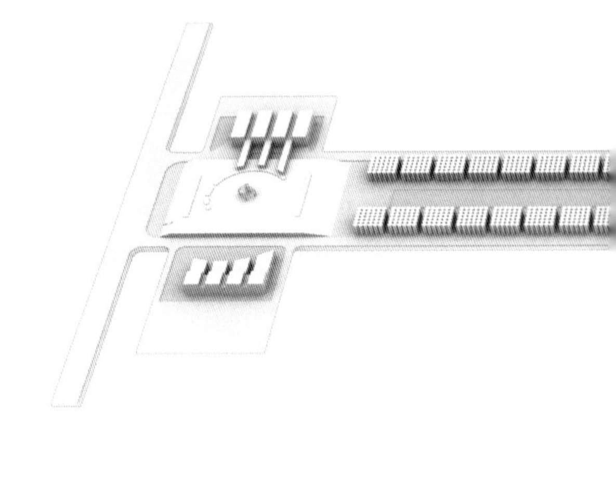

中华世纪门

China Arch for New Millennium

主题：传统更新版

Theme: Traditions Updated

中华世纪门选址在北京中轴线北端800公顷的森林公园内，南北长1300米、东西宽400米。在1300米的主轴上规划了青少年科技馆、艺术馆以及以“水”为主题的纪念空间——“水苑”；在600米的次轴上规划了以“木”为主题的“草院”；两轴交汇之处是主体造型——中华世纪门及现代科技城。

中华世纪门的主轴设计借鉴了北京天坛的布局特点。根据地段地形特点设置的副轴和斜轴，是对

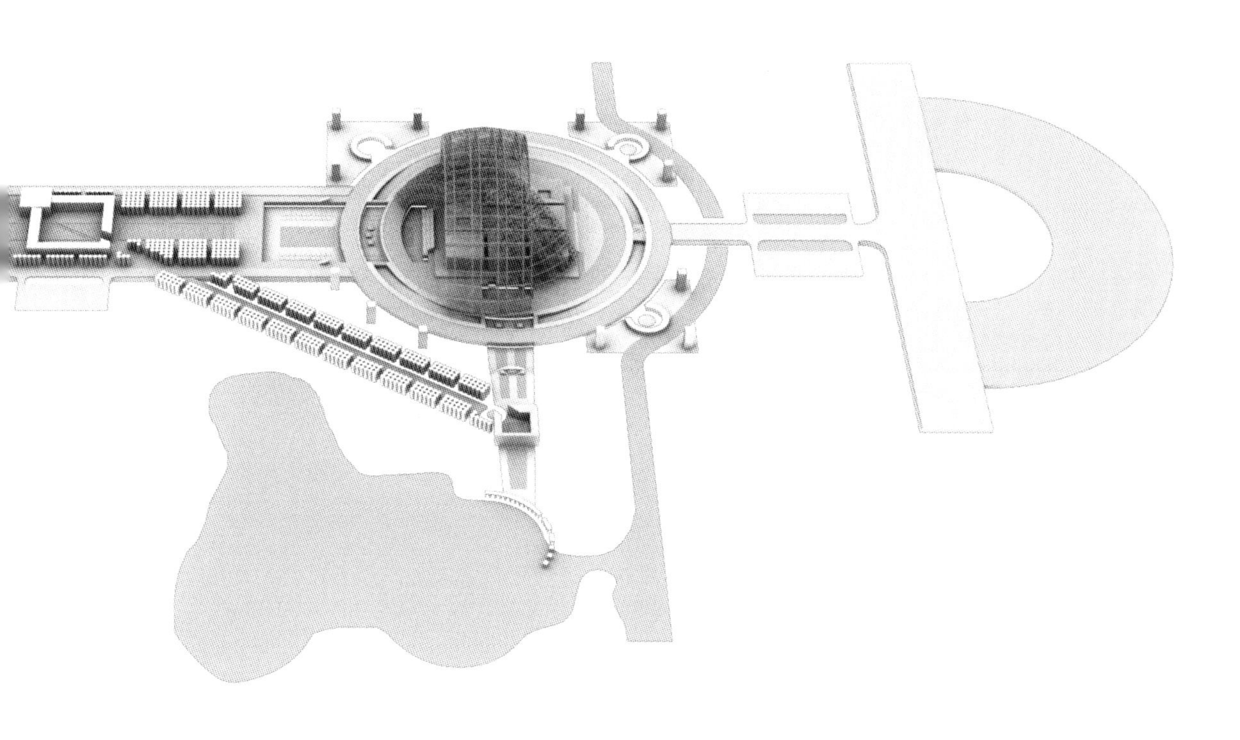

西式园林轴线设计手法的借用。三条轴线构成中华世纪门的规划骨架。

The China Arch for New Millennium is located in the northern end of Beijing's central axis, inside the forest park with surface area of 800 hectares; it measures about 1,300 meters from north to south, and 400 meters from east to west. On the 1,300 meters of central axis lays the Science and Technology Museum of The Youth, Art museum, and the "Water Garden" - a memorial area with water theme. On 600 meters of horizontal axis lays the wood-themed "Grass Yard". The main body of Chinese Arch of New Millennia and the Modern Science and Technology Museum is located by the intersection of the axes.

The axes design of the China Arch for New

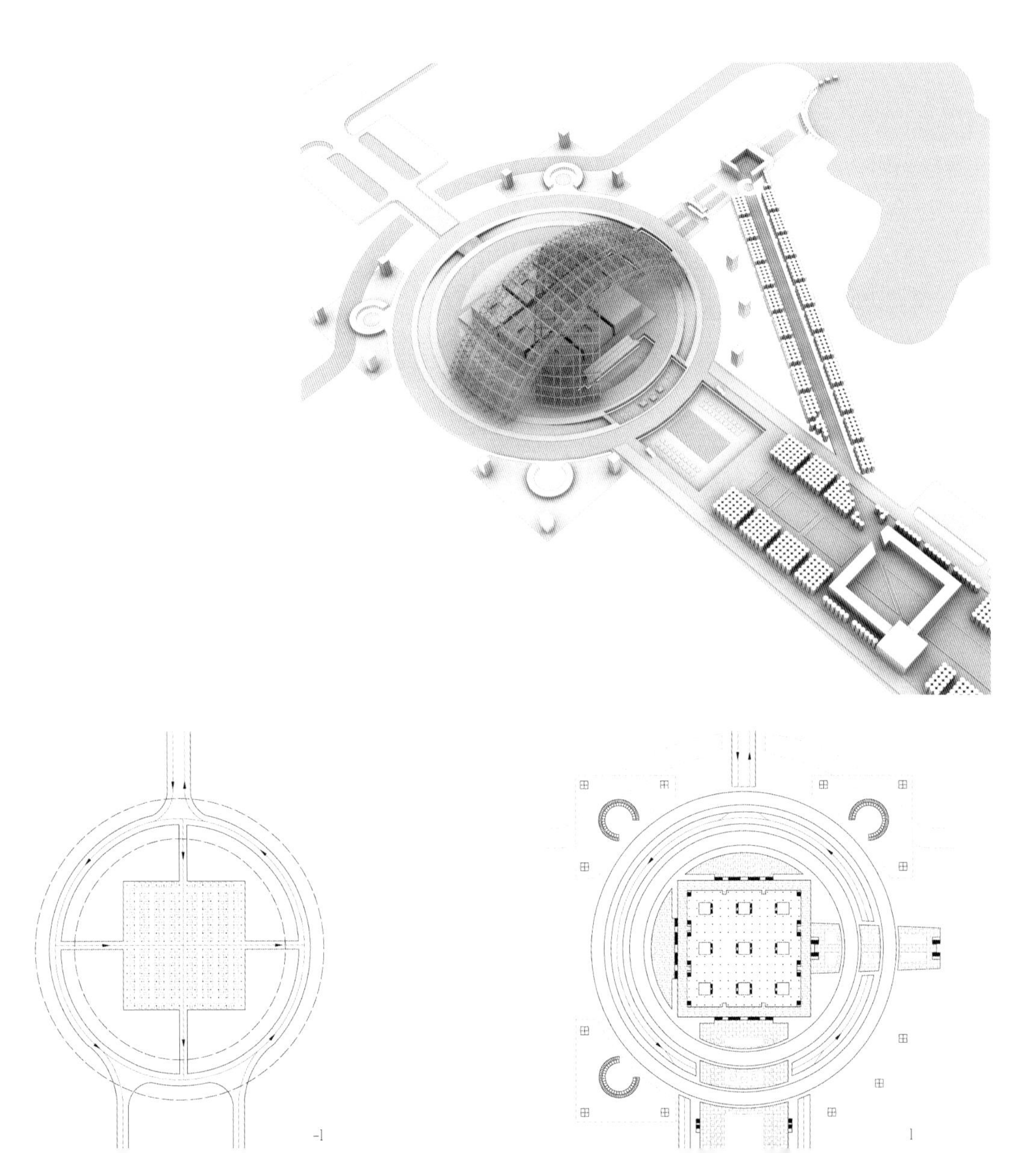
-1
1

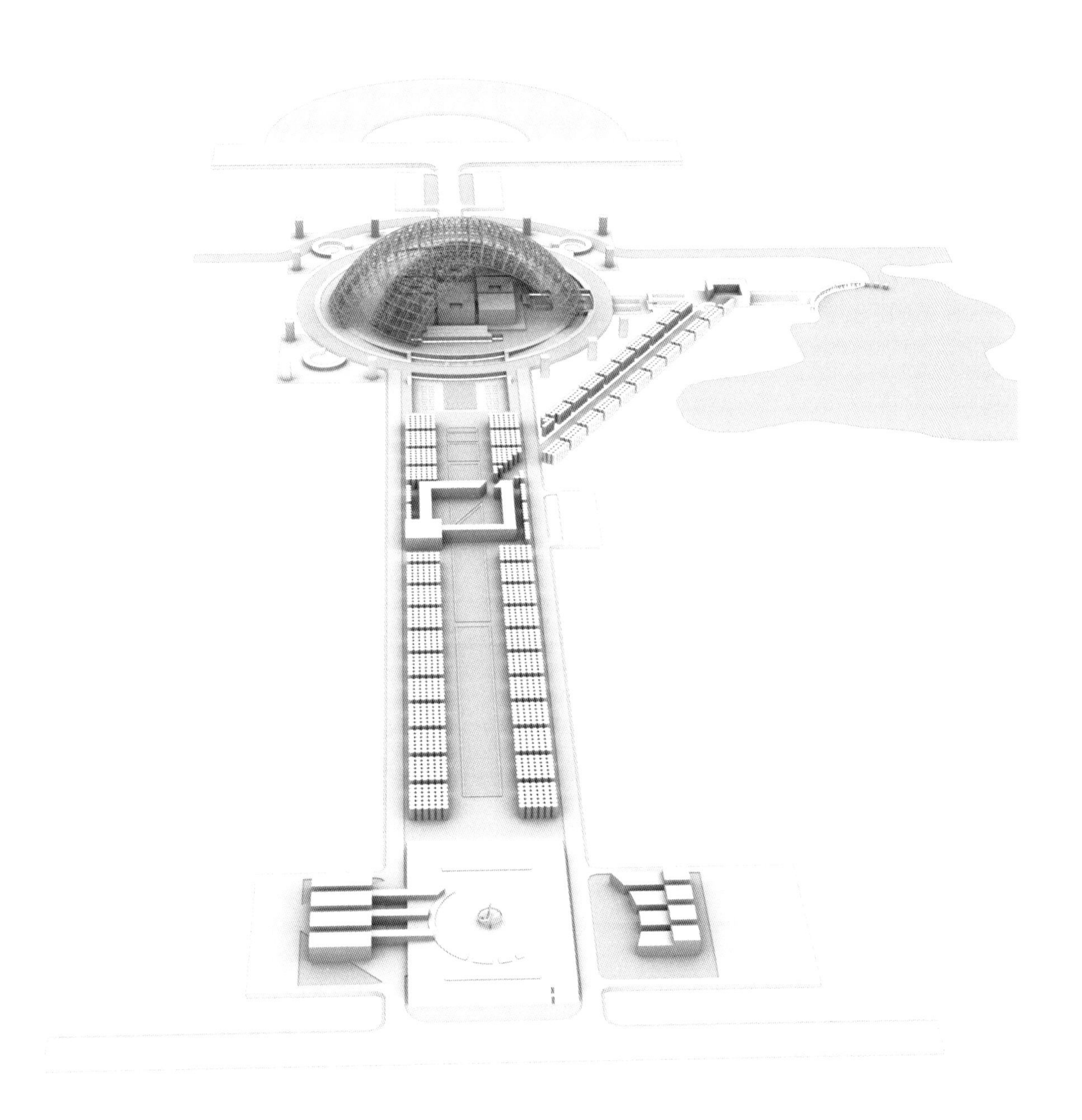

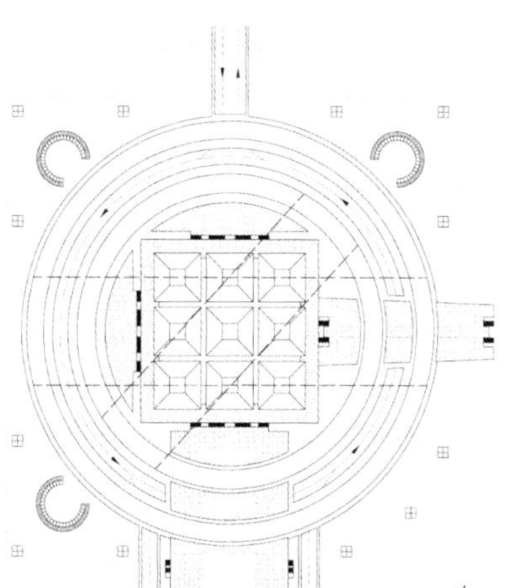

Millennium references the layout of Imperial Vault of Heaven in Beijing. The use of primary and secondary axes with according to geographic characteristics borrows western gardening design techniques. Three axes compose the framework of the China Arch for New Millennium.

曹妃甸 AB 地块概念设计

Conceptual Design of Caofeidian AB Plots

主题：打造绿甸方舟

Theme: Building the Ark of Green

曹妃甸 AB 地块是曹妃甸国际生态城的起步区。本方案在整体布局上引入新城市主义的理念，采用小街坊布局，并提倡土地混合使用，以保证街区的多样性。同时，倡导绿色生活方式，规划了城市大众运输系统以及城市慢行路网，鼓励步行与自行车等低能耗的交通工具。

另外，本方案在运河两岸设计了具有滨水特征的城市形象。一条沿运河由北向南直入大海，成为贯穿整个区域的景观轴线，串联起几个重要的城市

景观节点。运河两侧布置商业街，提供给居民商业和休闲场所；城市之门由办公楼和空中展览馆组成，是城市主题标志；绿甸方舟伸出海面，内部设置了以环保为主题的文化、娱乐及展览功能，提醒人们爱护地球环境。

Caofeidian AB block is the starting point of the Caofeidian International Ecology City. The project introduces the concept of urbanism in the

overall layout; it adopts layouts of small neighbourhoods and promotes mixed use of land to ensure diversity in the districts. At the same time, the city is built with public transportation system and promenade network to advocate green living. The city also encourages the use of low energy consuming transportation such as bicycle and promenade.

In addition, the project designed a waterfront city image on the shores of the canal. The canal runs to the ocean from north to south through the whole district, linking important city landscapes like a axis. Commercial streets on the side of the canal satisfy commercial and leisure needs for residents. The gate of the city, compose of commercial buildings and the Air Exhibition Centre, is the city landmark. The ark of greens reaches out to the ocean body,

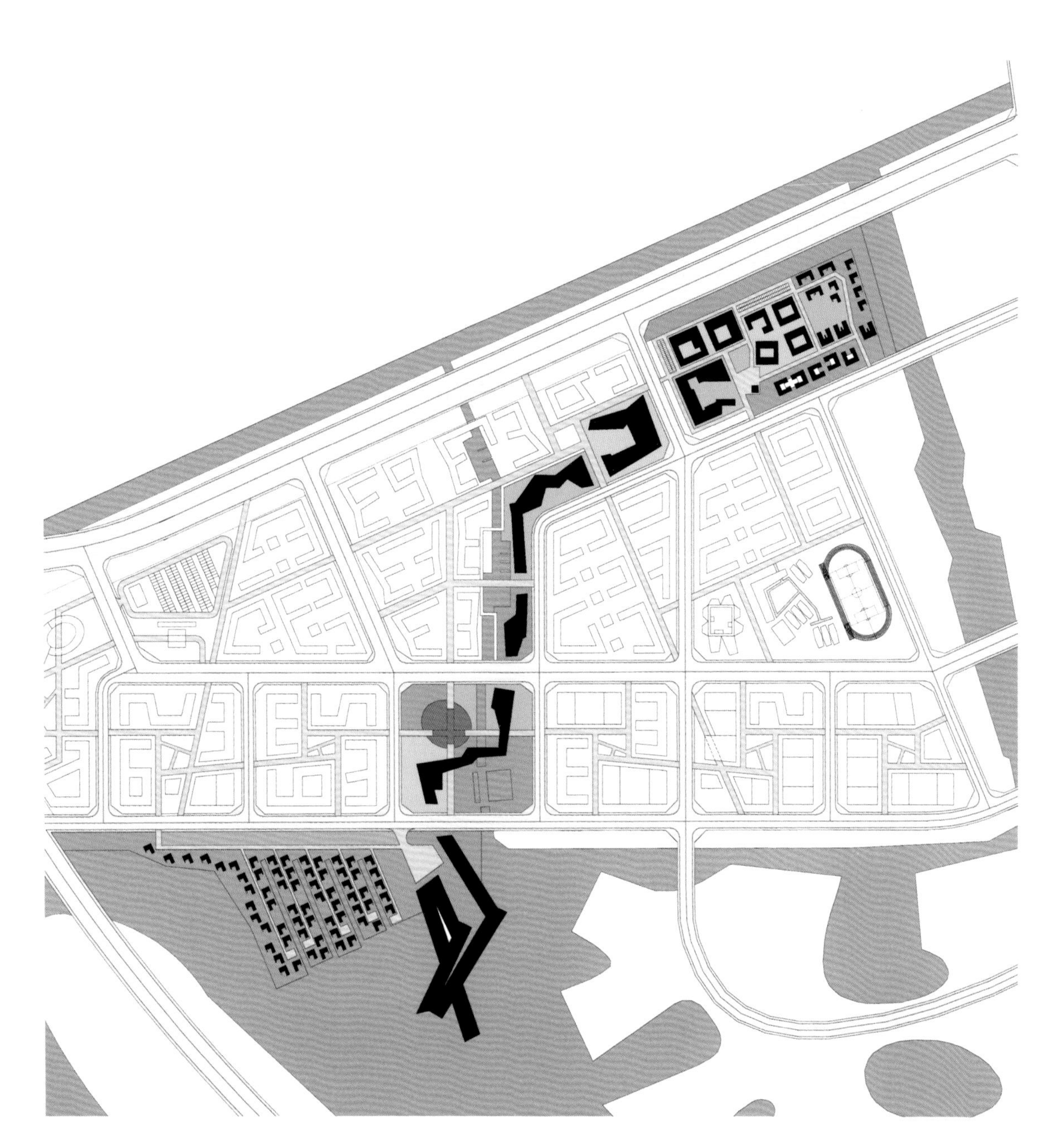

填海造地
Sea reclamation

remind the people to cherish earth's natural environment with the green-themed, cultural, and entertaining exhibitions inside.

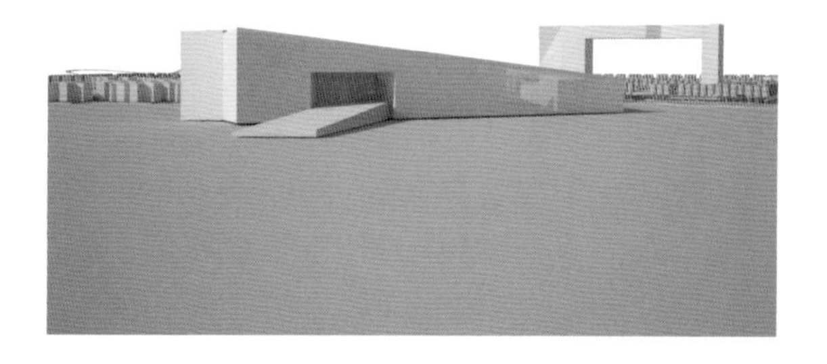

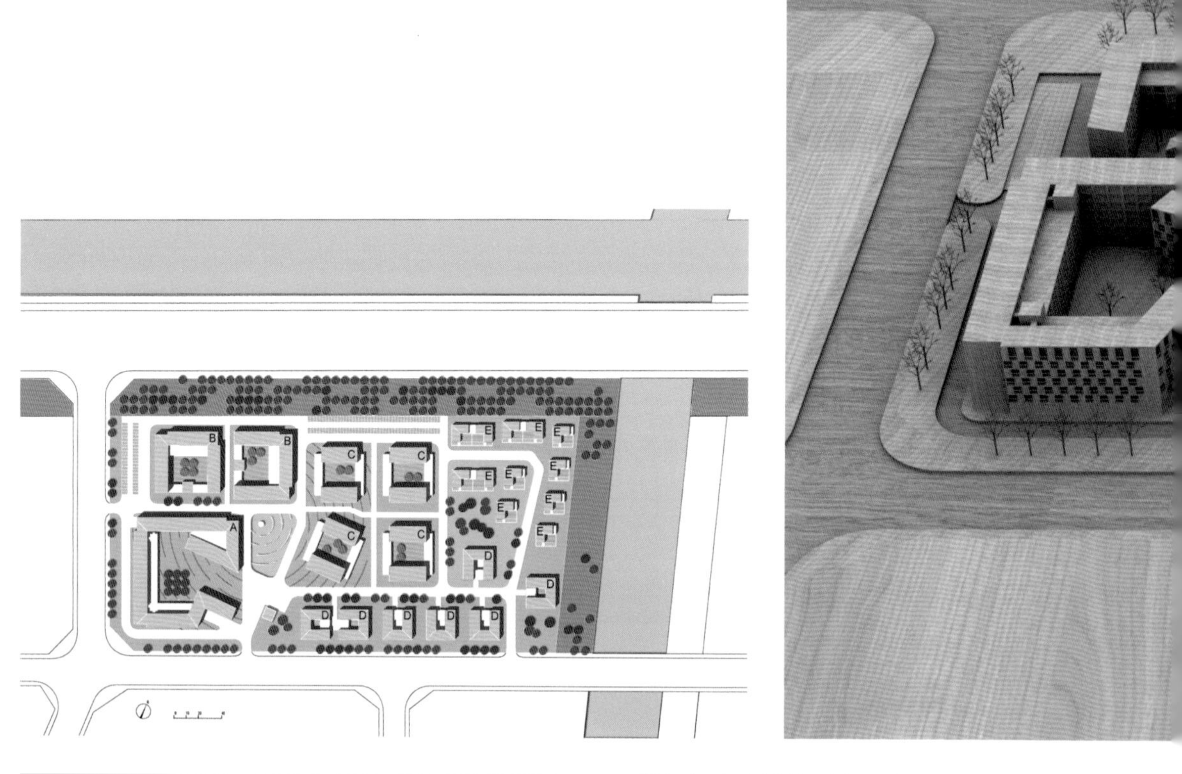

曹妃甸 B3 社区

Caofeidian B3 Community

札记：一个新型社区的尝试

Note: A Test in a New Form of Community

曹妃甸生态城位于唐山南部，是与曹妃甸工业区和港口配套发展的新城。生态城规划用地面积30平方公里，其中第一期建设的起步区为12平方公里。本方案B3社区是起步区中前期开发的社区，对生态城的发展具有重要指标意义。下面从合院式邻里结构、混合式功能布局及适宜性生态策略三个方面对B3社区的规划设计理念加以介绍。

· 合院式邻里结构

合院式住宅是中国的传统住宅形式，广泛分布

在很多地区，例如北京的四合院、云南的“一颗印”、徽州的“四水归堂”、福建的土楼等。这些居住形态不仅是当地地理气候的反映，也是当地人文传统的体现。中国自1950年代起，从前苏联引进行列式住宅模式，传统邻里结构被瓦解，并大量消失。而在一些西方城市，尤其是美国，从第二次世界大战之后，居住郊区化进入快速发展过程，居民大规模迁往郊区，住宅形式千篇一律，公共建筑散置各处，城市通勤距离被拉大，并导致居住分异现象的出现。

1990年代起出现的新都市主义开始对居住大规模郊区化和传统邻里被大量破坏的现象进行反思。新都市主义主张取代向郊区蔓延的发展模式，回归具有邻里生活氛围的紧凑型（compact）社区；提倡社区内街道呈网络结构分布，增加社区的可步行性；提倡社区中心和边界要具备可识别性，用院落和街道来展示都市生活。

曹妃甸生态城B3社区的规划引入了上述新都市主义的理念，整体空间结构上采用了合院式布局。

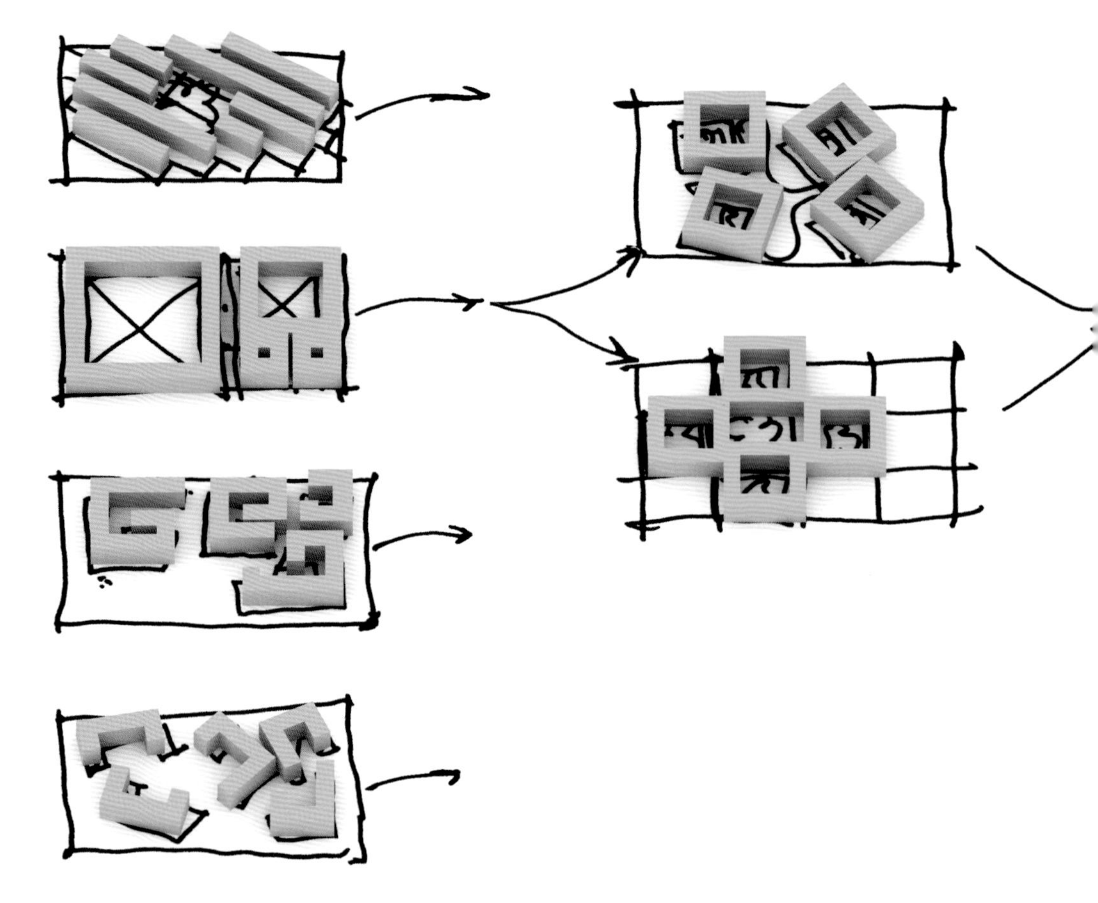

合院式布局不仅可以增加邻里交流机会，还有利于抵御曹妃甸冬季的大风。传统居住空间模式的引入，不是单单形式上的照搬，而是对传统居住空间模式的再创造，以适应当代社区生活的意识和步调。为此围合的院落做了适当开敞处理，让院落内空间和街道相贯通，形成多样和连续的公共场所，为展示邻里生活提供更多可能。

· 混合式功能布局

简·雅各布斯（Jane Jacobs）曾在《美国大城市的生与死》一书中对现代主义城市的功能性规划与发展方式进行了批判，指出过度强调单一的功能分区导致"睡城"和居住分异的出现。新都市主义理论重新审视了城市公共交通与土地使用模式的关系，提倡土地的混合使用（mix-use），即居住用地和其他用地的混合，不仅可以为居民提供配套的商业和服务设施，还可以为居民提供就近就业的机会。另外，不同类型的住宅混合在一起，可以有效减少居住分异现象；街道上不同功能类型的空间混合，可以为居民提供多样性的活动场所，增加人与人之间的交往机会；用地以不同规模和方式进行再次分割，由不同的地产商开发，可以避免形成封闭的大型独立社区。由此一个多元和混合的城市结构才能得以实现。

B3 社区在仅 8 公顷的用地内，实现了住宅、办公、商业和社会服务等不同功能类型的混合，为曹妃甸生态城提供了一个与以往完全不同的社区类型。B3 社区在功能混合上的尝试是多层次的，不仅体现在整个社区，还体现在建筑之间，甚至建筑内部等各个尺度上。住房依居住条件分为 5 个等级（单身宿舍、普通住宅、商住单元、高尚公寓、别墅），将不同阶层的居民混合纳入一个社区；而办公依空间类型亦是多种多样的（独栋、SOHO、共享），可以满足不同工作类型的需求。这个新型社区具有高度混合的功能，亲切宜人的尺度，激发互动的场所。其空间结构完整有序，同时容许在其中灵活植入新的功能组成，以适应多种多样的变化和需求。

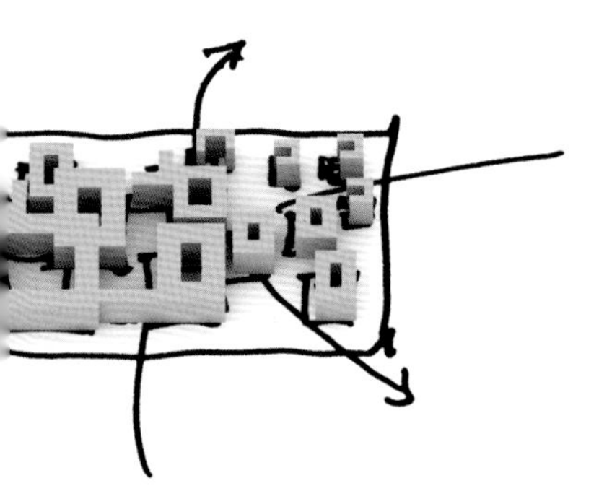

· 适宜性生态策略

生态和可持续发展是当今城市发展的一个主题。从社区规划层面上讲，生态策略意味着为社区居民营造宜居的生活环境，同时将能耗降低到最小程度。曹妃甸冬季海风强劲且非常寒冷，风车形的建筑组团布局和合院式的建筑形式，有利于防风并形成舒适的社区微气候环境；专用的步行道及自行车慢行系统的设置，便于居民采用环保节能的出行方式；不同建筑类型的混合布置，促进更加有效地利用空间资源和节约时间。

在技术层面上，B3 社区的生态策略可分为 5 个系统。社区内外资源体系应接系统——社区供电可实现与远期近海风力发电对接，供热可实现与远期工业余热供应对接；新能源复合应用系统——采用风光互补型路灯，太阳能供应热水，低层建筑采用地源热泵提供冷热源，多层建筑采用空气源制冷；源分离与分散式污废水处理系统——住宅建筑采用污废水分离，废水与雨水统一处理后一部分进入景观水系，一部分回用于办公建筑冲厕，污水进入黑水处理站处理，经过后续湿地深度处理后，进入景观水系，景观水系作为区域浇灌用水水源；源分类的垃圾减量处置系统——住宅建筑厨房垃圾全部经过粉碎机粉碎后进入黑水收集系统，与黑水统一处理，其余垃圾分类收集，在垃圾处理中转站将可回收垃圾回收；社区自成体系的景观生态系统——社区污水深度处理所采用的人工湿地，同时兼有景观观赏性，雨水则以景观式生态草沟收集，同时过滤拦污，改善雨水水质。

绿色“硬件”的建设需要绿色“软件”建设的配合，如果居民不对生活垃圾进行分类，再高标准的废弃物处置系统也难以发挥作用。所以，除技术以外，管理层面上的生态策略亦非常重要。社会舆论要对居民绿色意识加以引导，对居民的绿色生活方式加以规范，让绿色行为成为生活的必需、甚至生活的时尚。

原文刊于《世界建筑》2010 年第一期，第二作者：史逸。

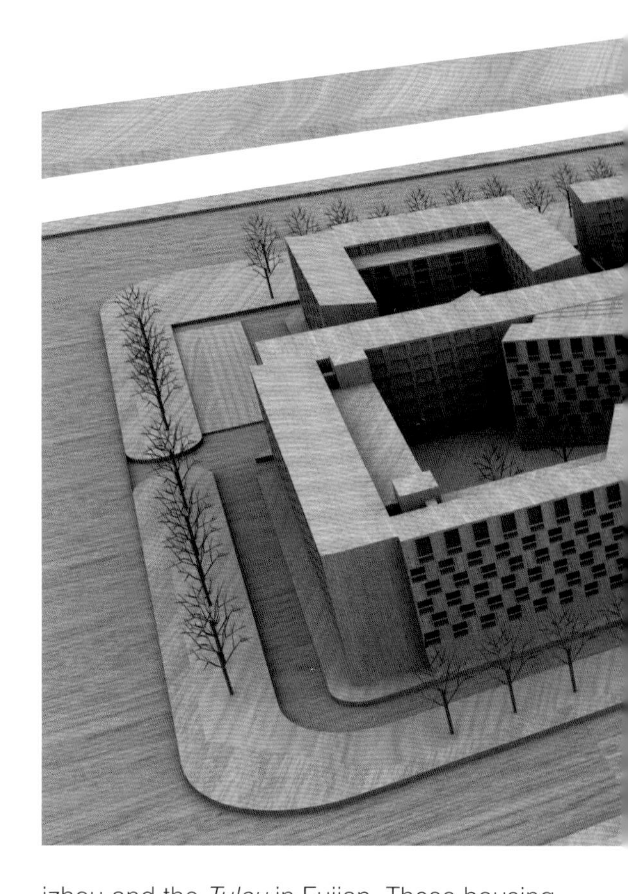

Caofeidian Eco-City, located in the south of Tangshan, is a new city developed in support of the Caofeidian Industrial Zone and port. The Eco-City is planned to cover 30 square kilometers, of which the first phase covers 12 square kilometers. Being one amongst the first stages in this phase, the B3 community is of great significance to the further development of the Eco-City. This essay introduces the planning and design concept of the B3 community from three aspects: its courtyard neighbourhood structure,the mixed-use layout and eco-strategy for suitability.

· *Courtyard Neighbourhood Structure*

Courtyard housing is China's traditional housing typology, which is widely spread in many local types, such as the *Siheyuan* in Beijing, the *Yikeyin* in Yunnan, the *Sishuiguitang* in Huizhou and the *Tulou* in Fujian. These housing types show not only the local geography and climate but also the local humanistic tradition. Since China introduced the "work-unit estate" mode from the Soviet Union in the 1950s, the traditional neighbourhood structure was broken up and began to rapidly disappeared. In comparison, in some western cities, after the World War II ,especially in the US, residential suburbanization began developing rapidly. Residents moved to the suburb in large numbers; housing types were always the same; with only few public buildings scattered here and there; and the commute distance to work in the cities was enlarged. These led to the emergence of residential segregation.

In response to this, the New Urbanism movement born in the 1990s began to reflect

on the mass residential suburbanization and the mass destruction of traditional neighbourhood structures. New Urbanism insists that the compact community with the resulting neighbourhood living atmosphere should replace the development of spread out suburban segregation instead streets inside the community should form a network to increase the walkability of the community; the community centre and borders should be recognizable, and courtyards and streets should be used to demonstrate city life.

The planning of Caofeidian Eco-City B3 community take into account this concept of New Urbanism mentioned above. The whole spatial structure adopts the courtyard-style layout. This courtyard layout can enhance neighbourhood exchanges and withstand the strong winter winds in Caofeidian. The introduction of this traditional spatial typology is not a simple copy of form but a recreation of it, so as to adapt to the pace and concept of modern community life. Therefore, the enclosed courtyard opens slightly to let the yard connect to the street and form various continuous public places to offer more possibility for showing neighbourhood life.

· *Mixed-use Layout*

Jane Jacobs once criticized the modernistic city planning strategies in *The Death and Life of Great American Cities*. She pointed out that overly stressing the division into separate functional zones led to the appearance of dormitory towns and enhanced residential segregation. The New Urbanism theory reexamines the relations between the city's public transporta-

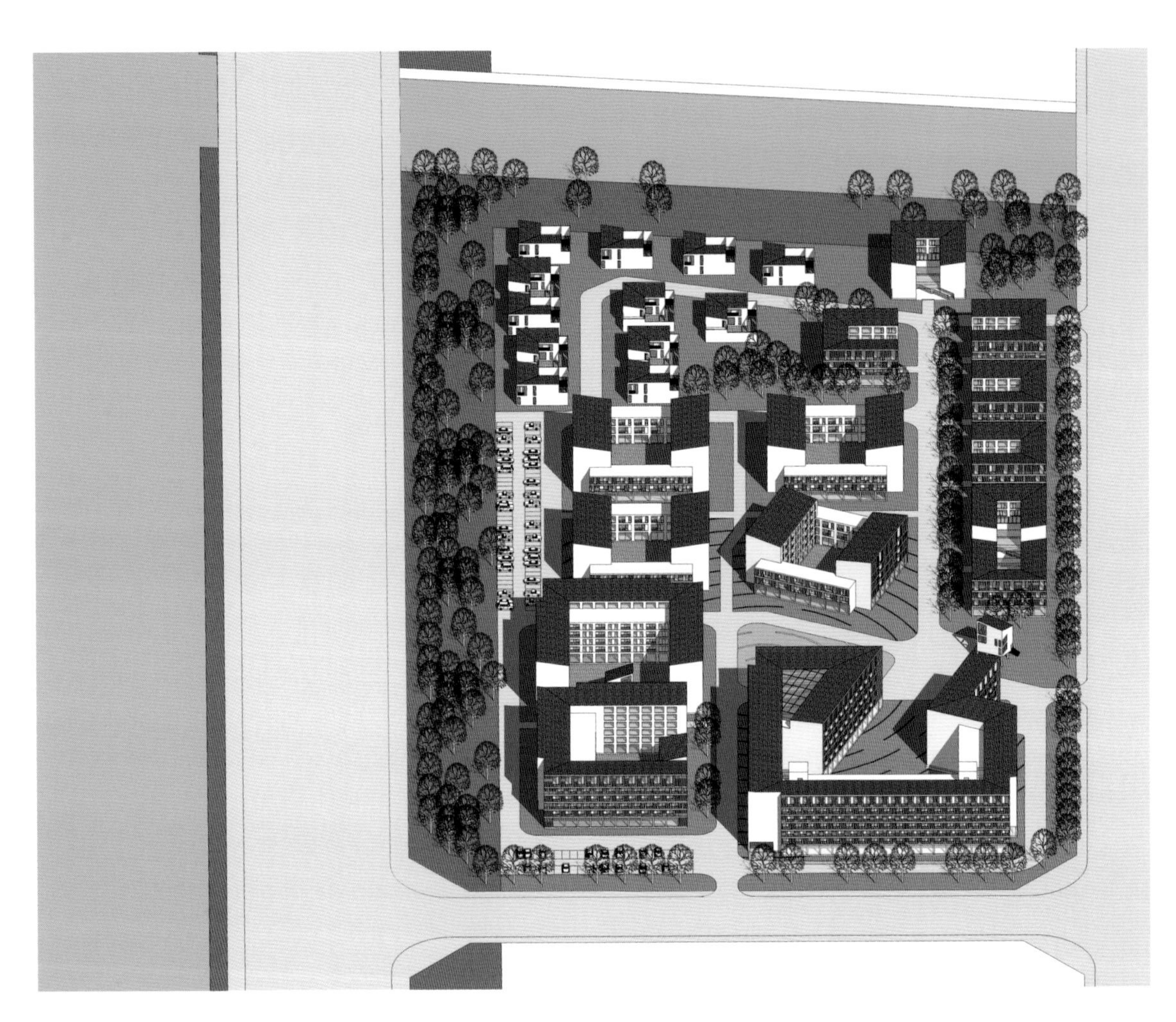

tion network and land use and promotes a mix-used land zoning, which means the mix of residential with related programs such as office and retail, which can offer not only supporting commercial and service facilities but also employment opportunities for residents. In addition, the combination of different housing types can effectively decrease the residential segregation phenomenon; as the combination

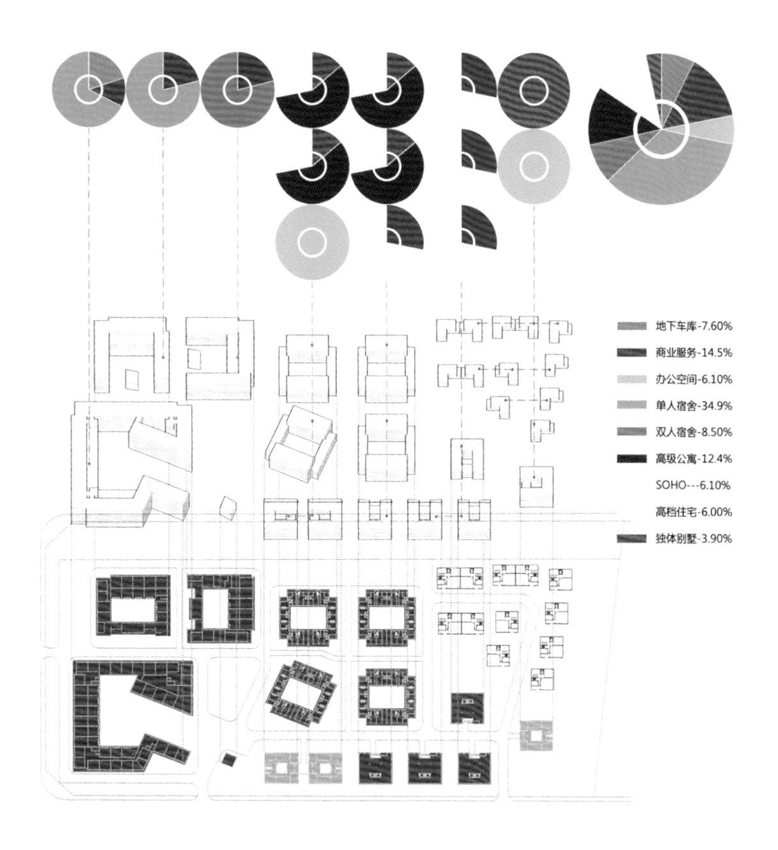

of spaces for various functions at street level can offer diversified activity places to residents and enhance interaction amongst people. At the same time, the process of land being subdivided into different scales and methods and developed by different land agents can also avoid the formation of enclosed large independent communities. In doing so a diversified urban structure can be formed.

Thus, in just 8 acres of land, the B3 community combines a mix of program, including residential, offices, commercial and social services, offering a totally different community type to Caofeidian Eco-City. The B3 community tries the combination of functions in various scales, which is shown not only on the scale of the community, but also inbetween buildings, and even in various dimensions inside

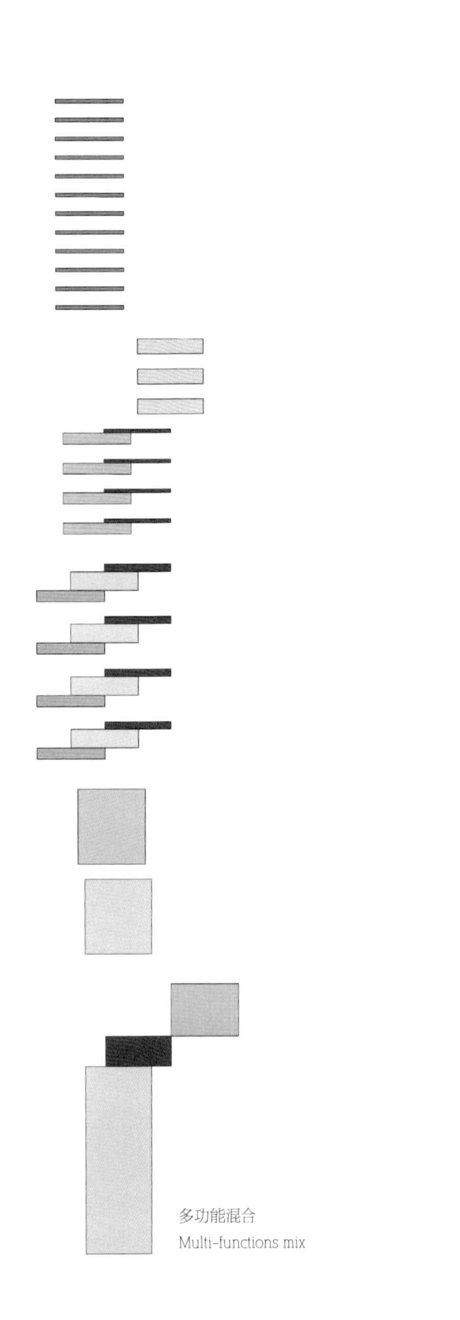
多功能混合
Multi-functions mix

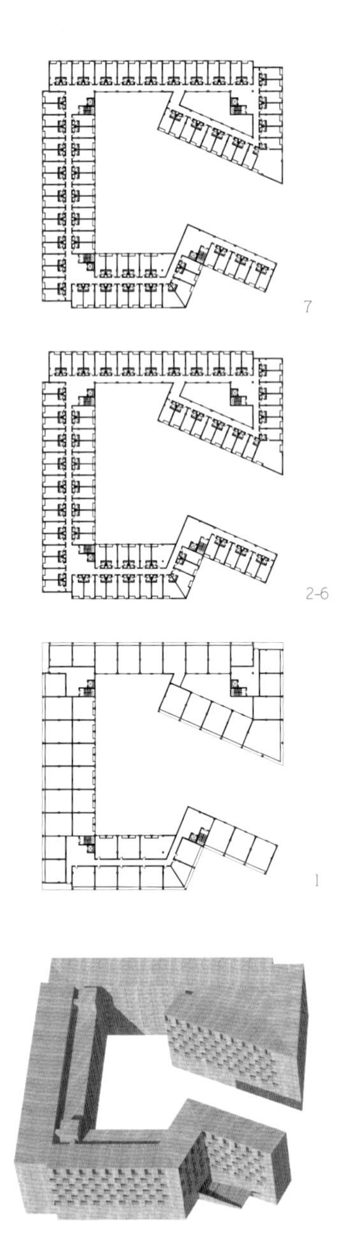

the buildings. According to living conditions, houses can be divided into five types (bachelor quarters, normal apartments, commercial apartments, noble apartments and villas), so as to incorporate various lifestyles into one community. The office spaces also accommodate various typologies (such as independent building, SOHO and sharing building) to satisfy the needs of different working types. This new

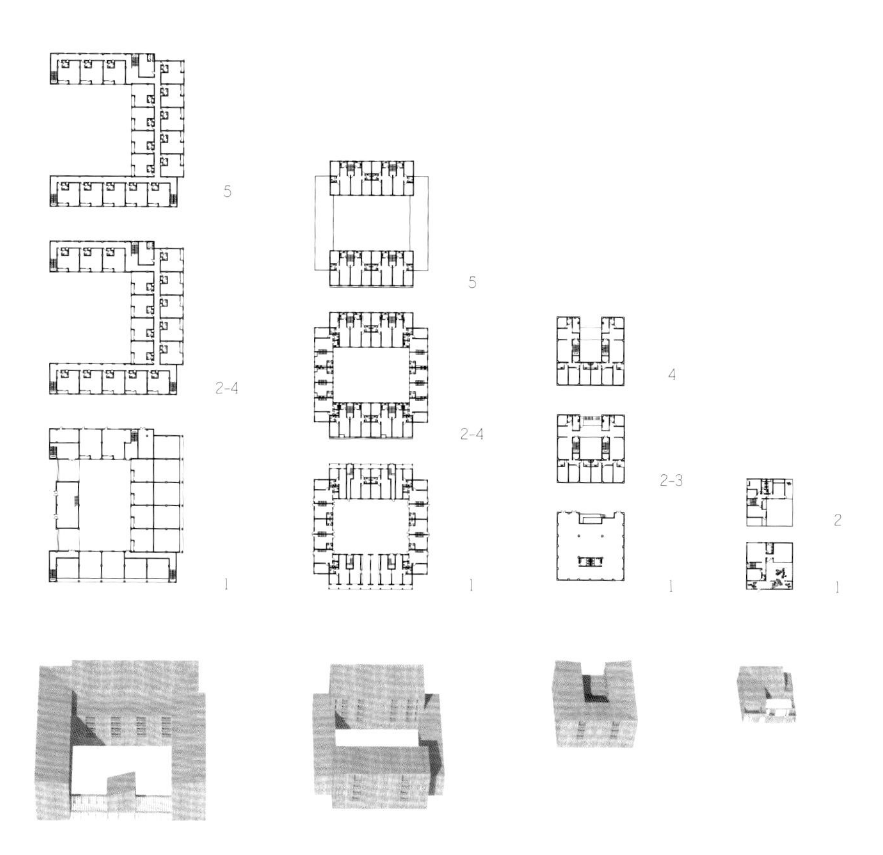

community thus boasts highly-mixed functions, agreeable dimensions and interaction-inspired places. The spatial structure is complete and orderly, and allows flexible integration of new functions, to adapt to various changes and demands.

· *Eco-Strategy for Sustainability*

Ecology and sustainable development are important themes in modern city development. From the community planning perspective, eco-strategy means to create a comfortable living environment for residents and to minimize the energy consumption. In Caofeidian, as the wind in winter is strong and cold, the layout of windmill-style building cluster and courtyard building types is beneficial to resist the wind and form an agreeable micro-climate environment inside the community; the special-

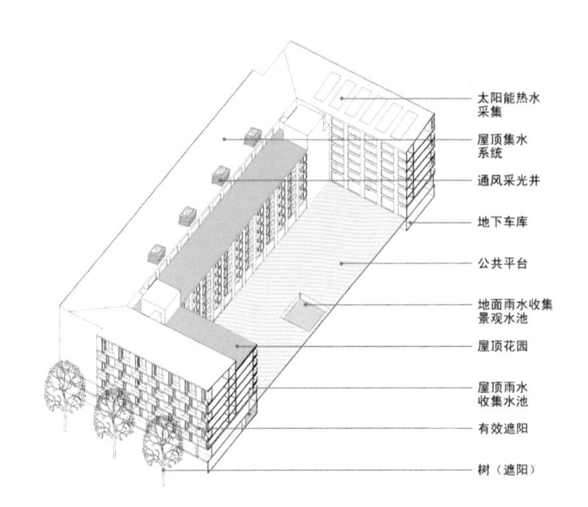

ized pavement and bike-slow-ride system are convenient for residents to adopt an environmentally-friendly and energy-saving transport mode; and the mixed setting of different building types can utilize space resources and save time more efficiently.

Technically, the eco-strategy of the B3 community can be divided into five systems. First of all the community's internal-external resource connection system in which the community power supply can be connected with the long-term offshore wind power generation. Secondly, the new energy composite application system in which wind-solar powered street lights are adopted, solar power is used to supply hot water, low-rise buildings adopts ground source heat pumps to for cooling and heating sources and multi-stories buildings use air based cooling systems. Thirdly, the decentralized treatment system of grey and black water: residential building separates grey and black water; after treated uniformly with rainwater, grey water will either be used in the landscape water system or to flush toilets in the office building. After collection, the black water will first enter a treatment station, and will further receive advanced treatment in the wetland, then enter into the landscape water system; and landscape water system is used as a regional irrigation water source. Fourthly, a garbage source-separation and reduction system in which kitchen garbage of residential buildings will be crushed completely by grinder and enter into the black water collection system to be treated with black water together; other garbage will be separately collected and

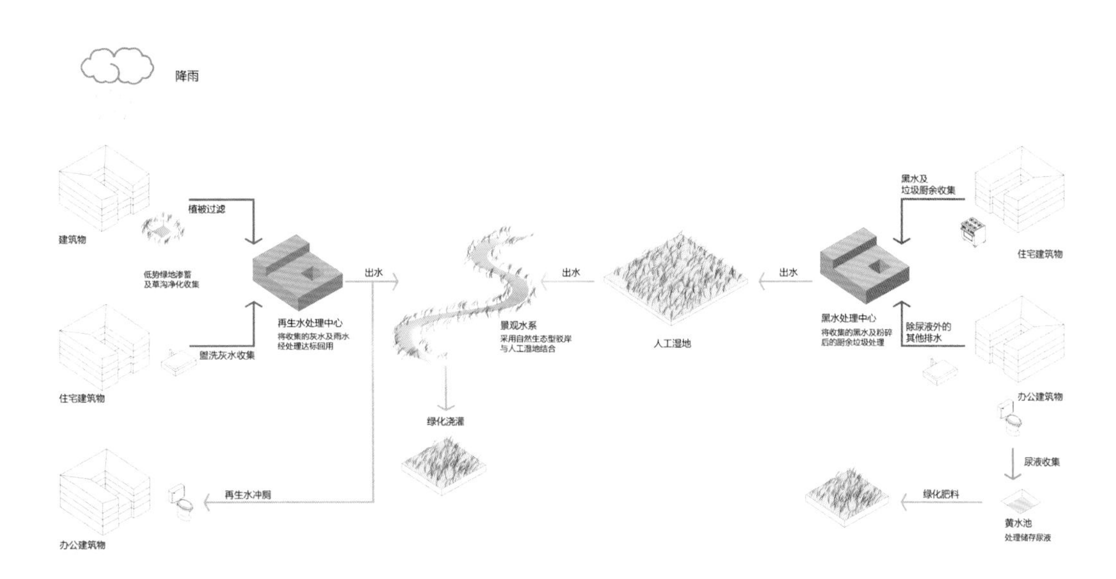

the recoverable garbage will be recycled in the garbage treatment and transfer station. Lastly is the community's self-contained landscape ecosystem: the advanced treatment of community sewage will use the manmade wetland which also provides beautiful landscape scenery; and the rainwater will be collected through a landscape bio-swale and be filtered to improve the quality of the collected rainwater.

The construction of green "hardware" needs the support of the construction of green "software". If residents' do not classify their household garbage, no matter how high the standard of waste disposal system is, it won't work. Therefore, besides technology, the eco-strategy is also of great significance from a management perspective. Public opinion should guide residents' green consciousness and regulate residents' green life style, to let green behaviour become a conscious part of life, a kind of culture and a kind of fashion.

Originally published in *World Architecture*, 2010 (01), with the second author Shi Yi

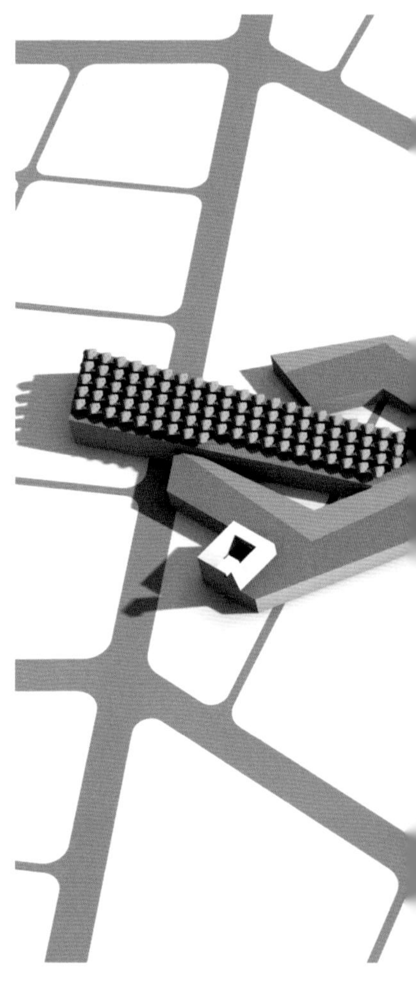

呼伦贝尔新区概念设计
Conceptual Design of Hulun Buir New District

札记：城市作为第二自然

Note: City as Artificial Nature

早在1872年，恩格斯在其《住房问题》一书中，就已提出“第二自然”的观念。德国哲学家黑格尔在更早之前也有类似看法：“第二自然”即设计自然（design nature）。自然是一种社会建构，不能脱离人类社会。恩格斯强调人为的自然，唯有透过“内在的相互关系”（inner interconnections）来理解。

近年来，传统的生态学应用已不限于“纯粹的”自然环境，如森林、湿地、溪流等，而是向城市及空间化的方向发展，即以“生态学的空间化”（spatial

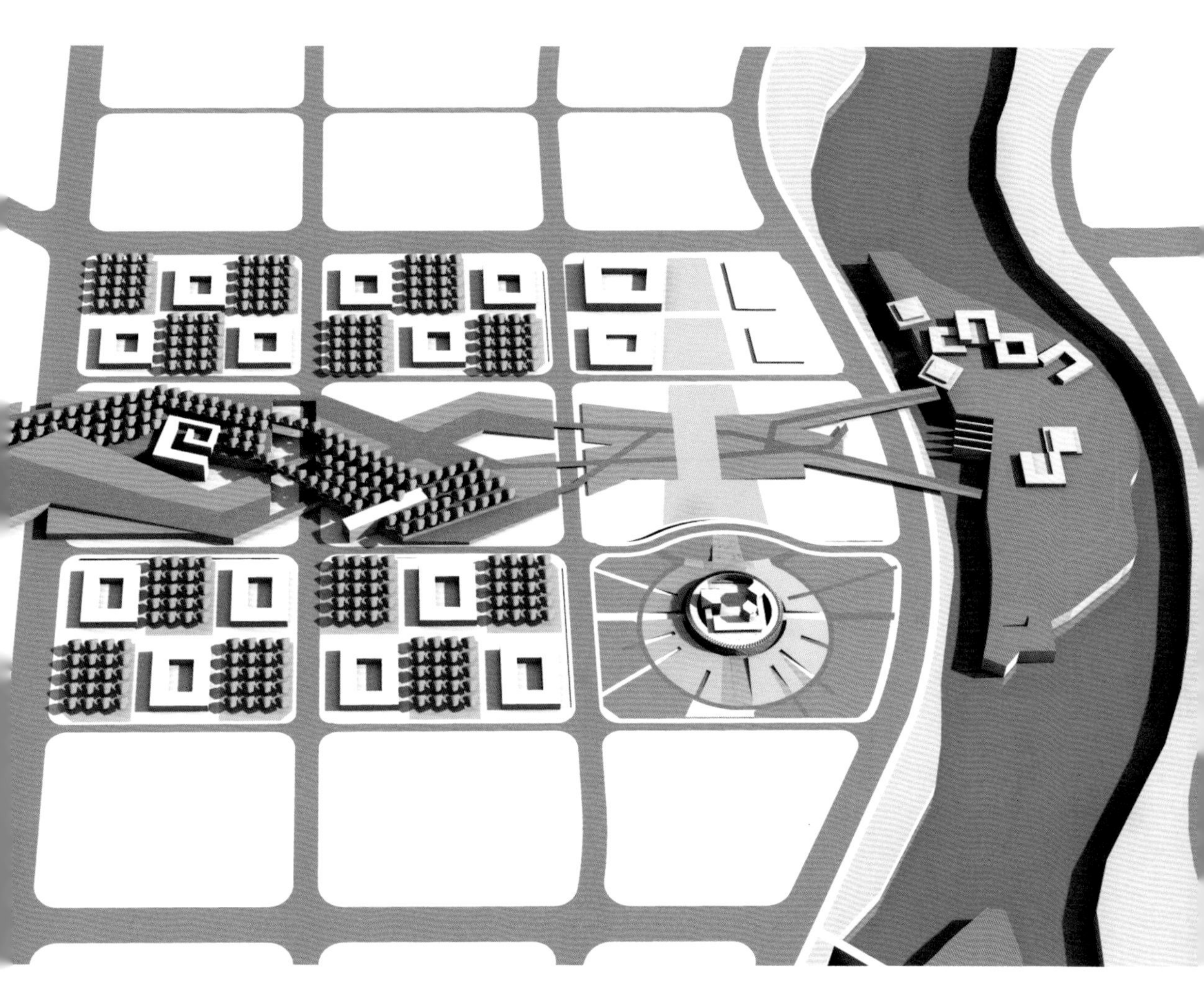

ecology）来研究受到人为影响与改变下的环境系统之形式、功能运作与空间模式，及其与生物、物质与能源流动之间的关系。另一方面，城市规划与设计也正迎来新的思维，向生态化靠拢。生态化的城市区域被看成“第二自然”，城市设计可以作为一种生态的介入（urban design as an ecological intervention），将改变自然环境系统与城市空间形式的关系视为一种设计的目标，引导空间演变的过程。

在呼伦贝尔新区的概念性城市设计中，我们借鉴了“第二自然”的理念，并在以下3个层面上，对新区的生态化设计进行了探索。第一，考虑地域自然条件下的太阳能接受度，以及防风等关键性的自然力与因子，推导出反映自然特性的城市空间基本结构；第二，模拟草原地貌、河流、植被等形态特征，构建反映地理环境的城市景观系统；第三，吸收地域民族文化，提炼体现地域文脉的建筑模式语言。

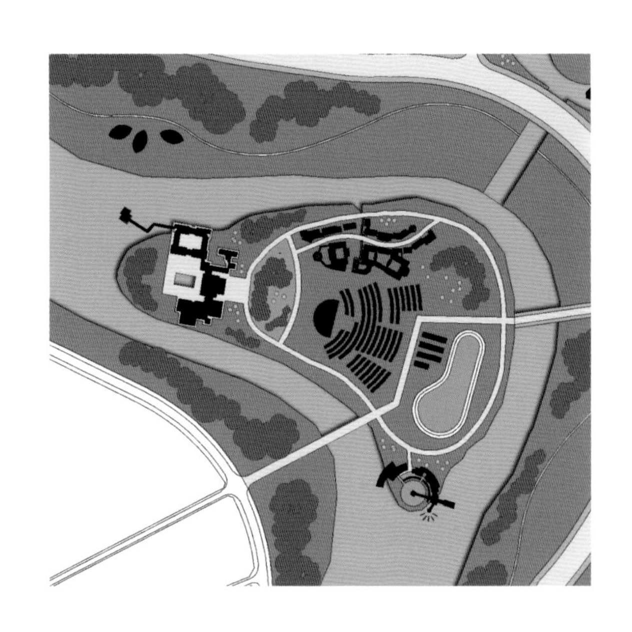

· 反应自然特性的城市空间结构

太阳能接受度和强风预防措施是呼伦贝尔这种草原型城市设计首要考虑的生态策略。呼伦贝尔光照资源丰富，年日照时数平均 2805.2 小时，年日照百分率 63%。一个有效的被动太阳能吸热系统，使建筑在冬天可以接收到最大的太阳辐射，在夏天可以有效地防止太阳辐射。为了取得最佳朝向，设计团队计算在 360° 朝向中，每 5° 变化在单位面积里垂直面的太阳能的得失。每个朝向中得到三个数值：在最冷的 3 个月中接受到的太阳能，在最热的 3 个月中得到的太阳能，全年平均接受到的太阳能。建筑最佳朝向的布局原则是根据冬天和夏天最佳朝向的平衡而得到的。因此，根据这种权衡而取得的最佳朝向，并不是指向能取得冬季最大太阳得热的朝向，而是向东稍微偏移，使得建筑在夏天下午的时候受到保护，而接受不到太阳的辐射。

呼伦贝尔新区分东山台地上、下两部分。草原型城市形态无天然屏障的自然因子，使得风向分析

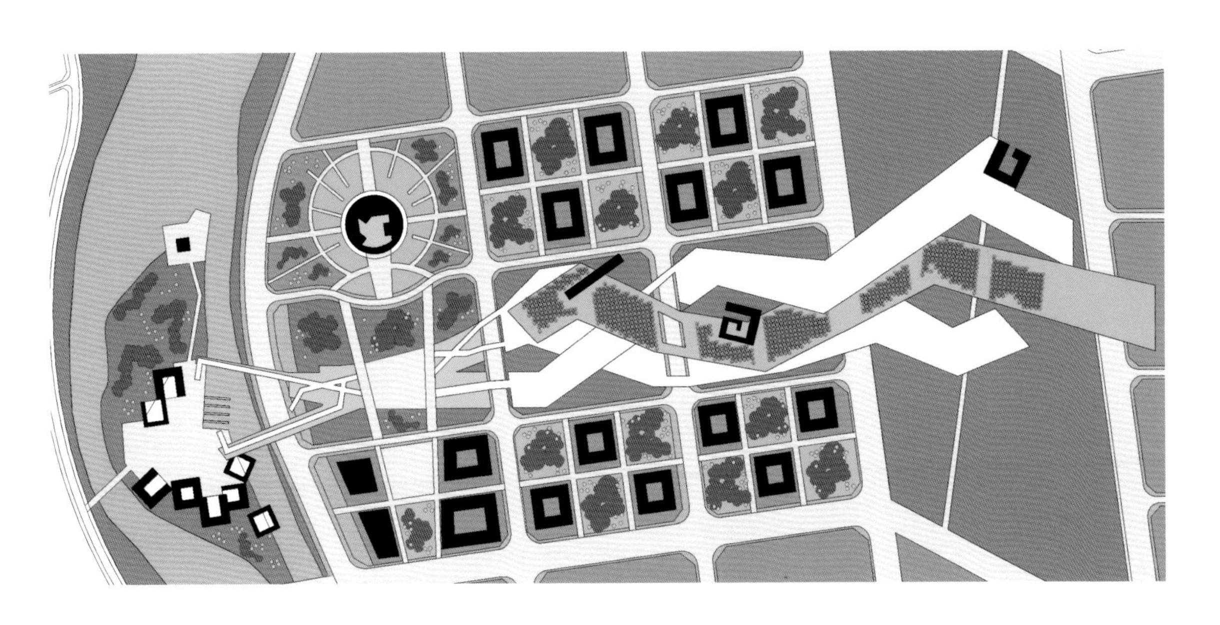

及强风预防策略很重要，对于台地之上的部分尤为重要。风向数据一般难以明确量化和显示，这是因为风向和风速经常变化。因此现实风向和风速须要考虑到时间和频率。设计团队测算用的风向系统，可以展示和分析风向数据。方向系统分析把风向数据分成16个不同的方向，以不同的时间间隔为单位进行分析。通过分析可以看出，呼伦贝尔在春季风速最大，但是并无强烈的主导风向，主要以南风和西南方向偏多。据此设计团队制定了台地上的防风策略，在城市内部设置防风林体系，城市空间结构的布局将防风林体系有机地纳入其中。

针对台地上和台地下气候条件的不同，我们选取两套网格来生成城市空间的基本形态，满足太阳能接受量及防风策略等生态要求。台地下区域基本延续了原有城区的网格机理，台地上区域除了延续了原有城区的网格机理外，另外置入一套正南北方向的矩形网格。两层网格配置既可以与原有城市肌理无缝对接，又能够保证台地上区域内建筑有良好的朝向。在两个网格系统交错形成的边角空间设置防风林地，一方面可以防止强风，另一方面可以形成新区良好的生态环境，激发城市活力。在草原型城市空间内部设置防护林的做法目前还不多见。这里大自然与人工互为镶嵌，交织在一起，构成具有“第二自然”特性的城市空间构架。

· 模拟地理环境的城市景观系统

城市设计并非是简单地寻求一种表面的美。城市设计在某种意义上是寻求一种包含人及人赖以生存的自然和社会在内的，以协调性、舒适性为特征的城市景观体系。城市设计关注的应该是城市乃至更大范围的自然生态和人工系统的统筹协调，落实一种建立在人类与自然共生基础上的综合景观环境体系。其设计原则是以自然地貌为依据，通过城市景观系统的介入，对原有自然景观进行优化和提升。

呼伦贝尔具有典型的草原地理环境特征，草甸草原、台地草原、河套湿地兼具。新区城市景观系统的设计力图在城市空间中模拟和优化这种特征，

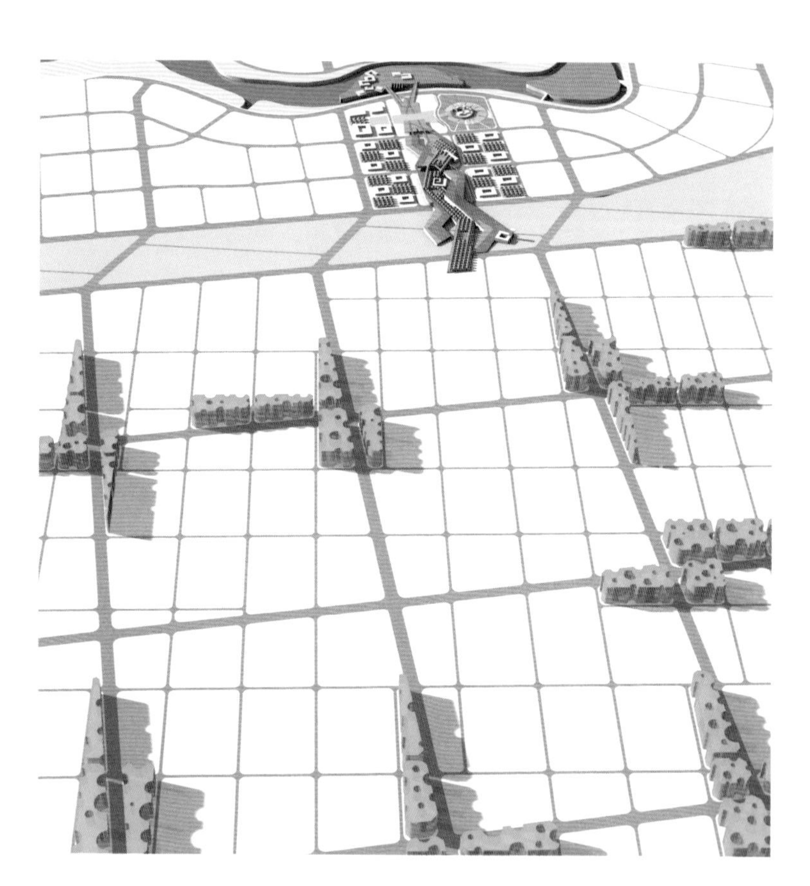

创造具有草原型城市美学价值的景观系统。

在呼伦贝尔新区，东山台地犹如一面城市背景矗立在东侧，伊敏河好似一条飘带在西侧由南向北串流而过。东山台地和伊敏河是城市最大的自然景观资源，对城市的景观体系起着支撑作用。为了强化这一地理环境特征，我们构建了一个“T”字形的景观骨架：一是从伊敏河上的河心岛向东一直延伸到东山台地之上的生态绿轴；另一是沿伊敏河滨水岸边从北向南设置的滨水景观轴。

生态绿轴约有600米长，从伊敏河上的河心岛出发，一直延伸到东山台地之上。与一般城市绿化轴线不同的是，这个绿轴标高渐渐升起，它联系起城市最主要的两个自然景观元素，打通了东山台地和伊敏河之间的视觉通廊，对原有自然景观起到提升和加强的作用。生态绿轴两侧依次安排了行政中心和文化旅游设施，绿轴架空跨越城市道路，以降低道路对绿轴的分割和对绿轴亲水效果的影响，同时在架空空间底下为城市提供了停车场所。

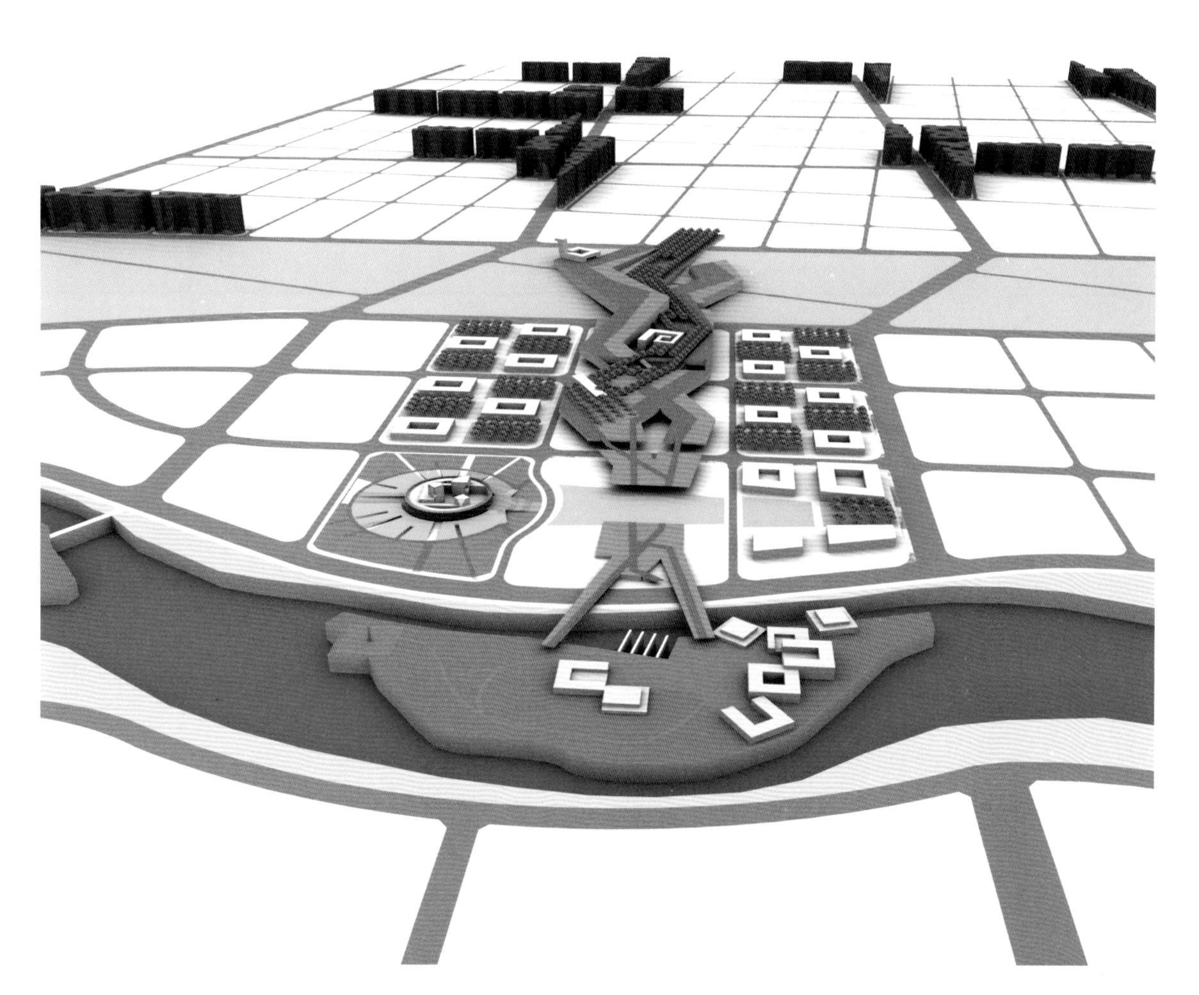

沿伊敏河岸线打造滨水景观轴。这个轴从北向南贯穿旅游服务区、行政办公区、商贸区和创意产业园区，在行政区与生态绿轴相交，局部放大形成广场。滨水道路两侧保留较宽的绿化带，在河岸内侧保留连续湿地生态带，以自然景观为主。在两个河心岛布置观光、休闲、游憩等功能，创造融入自然的城市生活环境。

· 体现地域特征的建筑模式语言

自芒福德(L. Mumford) 1947 年为了区别于国际式(International Style)提出地域现代主义(Regional Modernism)的概念以来，建筑的地域性渐渐引起人们的关注。尤其是 20 世纪末以来，在全球化、工业化的背景下，建筑地域性成为社会普遍关心的问题。从城市作为“第二自然”的角度看，建筑是“第二自然”中人工与自然的调节器，其不仅是人遮风避雨的庇护所，也影响着人的精神世界，它是特定时代的生活本质和社会特征。作为调节器的建筑更注重人对环境的感受和互动，更注重建筑

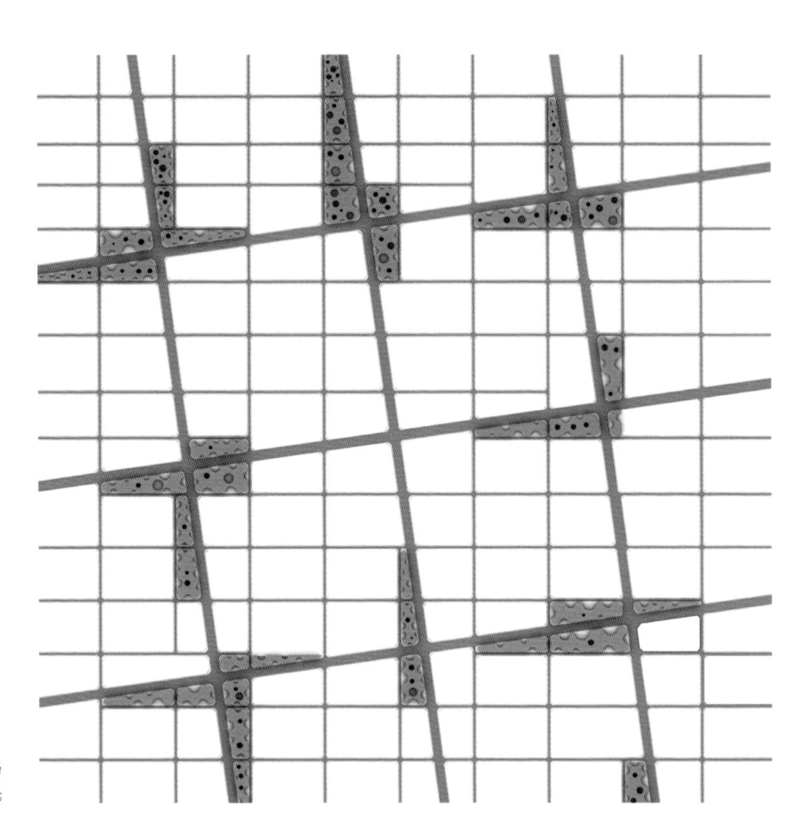

“大风车”防风林布局
Layout of Windbreaks

的此地此情此景，也就是建筑的在地性。在地性的建筑模式语言是多元的、综合的，与特定的地域环境和社会文化相关联，具有民族性和地方性的特征，也具有时代性的特征。

呼伦贝尔的典型地域建筑莫属蒙古包。蒙古包的最大特色是顺应自然、利用自然、装点自然，它不但对当地自然条件的适应性强，而且用地域特征的建筑文化全面协调了人与建筑、人与环境、人与自然的关系。当然，建筑的地域特征并不完全是环境气候决定论，文化因素在很大程度上也会施加影响。蒙古族的文化因素有其独到之处，其传统文化对宇宙的认识是“天圆地圆”，它同蒙古族的宇宙观、思想意识、审美以及生活方式紧密相关。“圆”的形态出现在蒙古族日常生活、经济生产、宗教信仰乃至军事阵列中。“圆”被人们神化和崇尚。“圆”的美是一种自满自足的美，是一种既无开端又无终止，代表永恒的美。“圆” 是蒙古族对美好事物的向往和追求，是宇宙、日月等自然现象在心灵深处的投射。

在呼伦贝尔新区的设计中，我们尝试着从蒙古族文化吸收灵感，提炼体现草原地域特征的建筑模式语言，并应用到城市景观中结点性标识性建筑当中去。以行政中心为例，行政中心作为重要的节点在整个城市空间结构中起着控制作用。我们努力提炼和运用具有地域特征的建筑模式语言，通过轮廓外形的连缀和内在意向的追求，最终达到一种和谐统一、完满美好的效果，以期建筑意象获得广泛的认同感、归属感和自豪感。需要同时强调的是，这种具有地域特征的建筑模式语言又是现代的，它展示了地域特色与现代化风格的结合，展示了一种新草原文化。

·结语

在生态成为社会普遍关注焦点的时代，建筑师需超越传统的设计思维。新的设计理念和创意仰赖分析的力量，依赖资讯分析与解码的结果来推衍出建筑与景观空间的新形式。城市和建筑的空间形式

防风林分析
Analysis of Windbreaks

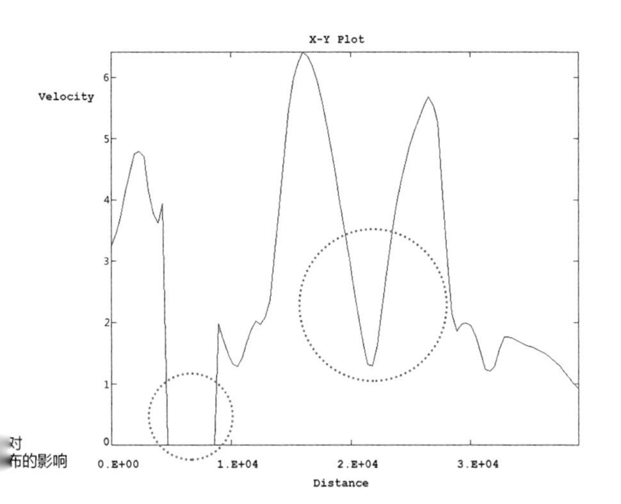

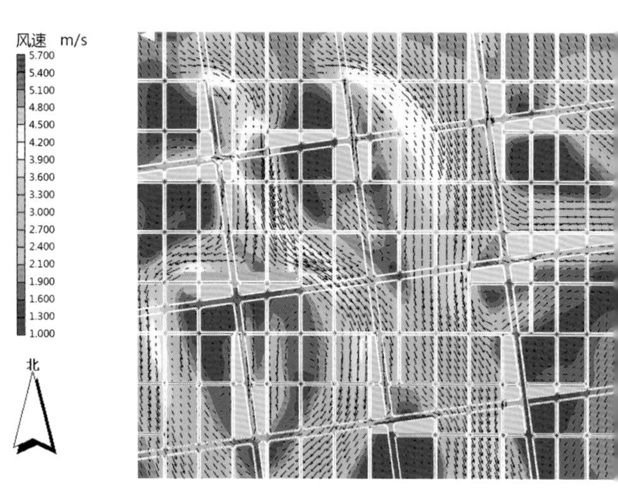

的演化需以精密的逻辑推演为基础，产生空间形式外在的社会、经济与政治力量，需与内在的空间设计逻辑等量齐观。普林斯顿建筑学院院长史坦艾伦(S. Allen)认为，设计犹如一门人造生态学(artificial ecology)，设计活动正如处理资源、物种与气候之间的复杂互动关系。建筑不再只是独立于基地上的人造物，亦即建筑与景观之间的边界逐渐模糊。城市作为“第二自然”，其规划设计需从环境中发掘机会与品质，打破建筑与基础建设(structure and infrastructure)，建筑与景观建筑(architecture and landscape architecture)、城市与景观(city and landscape)等传统分界之间的藩篱，从中寻求城市环境整合的创新方案，凝聚出具有实践能力的生态城市设计方法。

原文刊于《世界建筑》2010年第九期，第二作者：杨沛儒。

As early as 1872, Frederick Engels, in his work “The Housing Question”, had put forward the idea of an “artificial nature”. His predecessor, Hegel, had a similar view: that “artificial nature” essentially meant designed nature. Nature is a social construct, not to be separated from human society. Engels emphasized that human nature is only understandable through inner interconnections.

In recent years, traditional ecological applications have not been limited to examples of “pure” natural environments such as forests, wetlands, streams, etc., but have also tended towards the direction of urban and spatial development. That is, spatial ecology which is meant to study the impact of human influences and behaviours upon environmental systems, study functional and

spatial patterns, as well as study the interrelated flows of biology, materiality, and energy. Simultaneously, urban planning and design have also began to usher in a new way of thinking which approaches an ecological approach. Ecological urban areas can be seen as an artificial nature, and urban design can be seen as an ecological intervention. The main design goal revolves around the ability to change the relationship between ecological systems and urban spatial behaviours, thereby affecting the process of spatial evolution.

In the new conceptual design for Hulun Buir City, the idea of "artificial nature" has been heavily called upon. The following three ecological-design concepts are explored in this project: regional and natural condi-

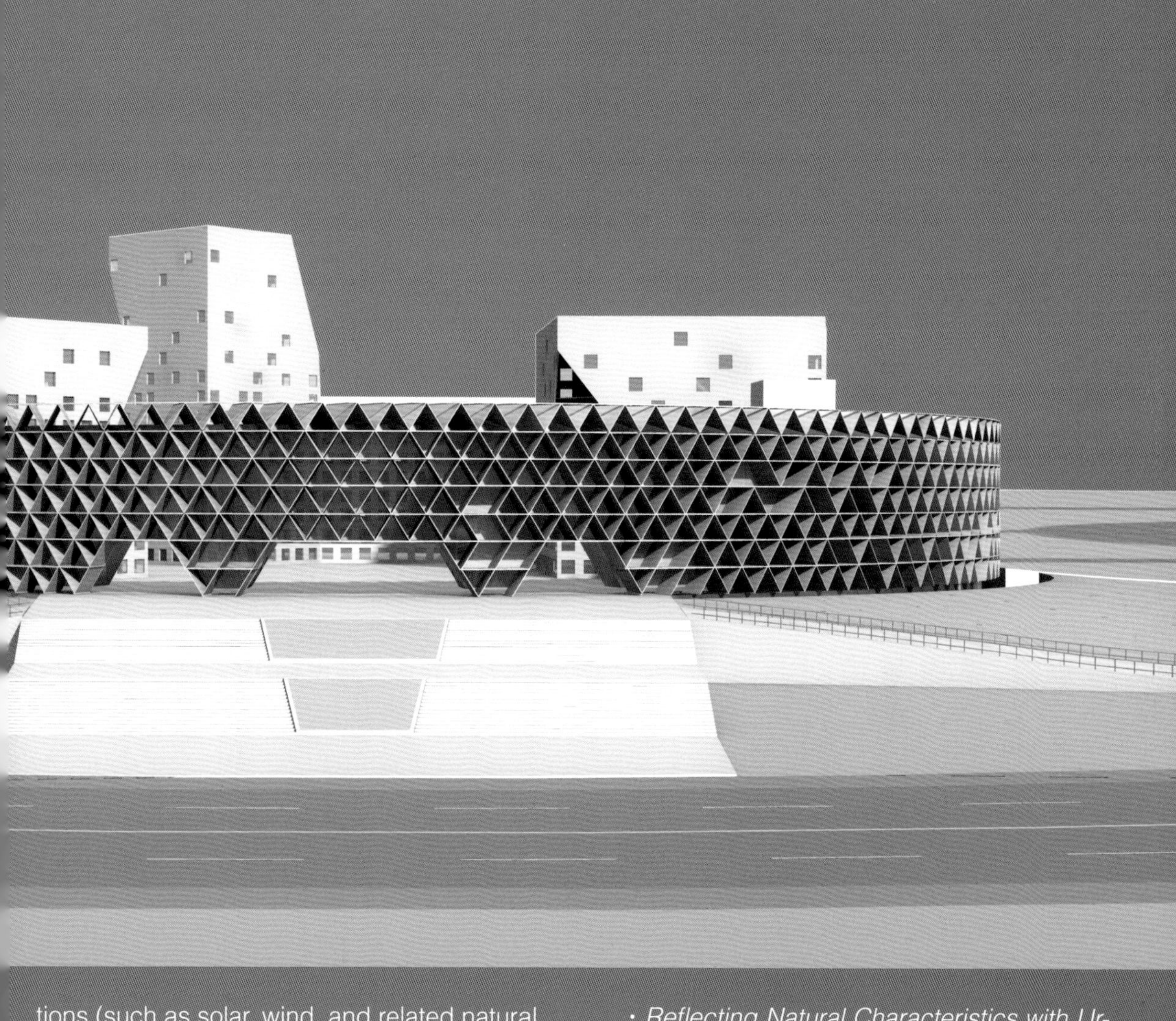

tions (such as solar, wind, and related natural forces) from which the basic structure of urban space will be derived; the simulation of prairie landscapes, rivers, vegetation, and other morphological characteristics which will accurately reflect the geographical environment of the city-system; the absorption of regional and local culture to embody a refined architectural language.

· *Reflecting Natural Characteristics with Urban Spatial Structure*

The primary ecological urban design strategies of the Hulu Buir design take into consideration extensive solar and strong wind factors on the Hulu Buir Steppe. Hulun Buir is rich in sunlight, it averages 2805.2 hours of sunlight a year (about 63% of the year is spent in sunlight). Valid passive solar heating systems

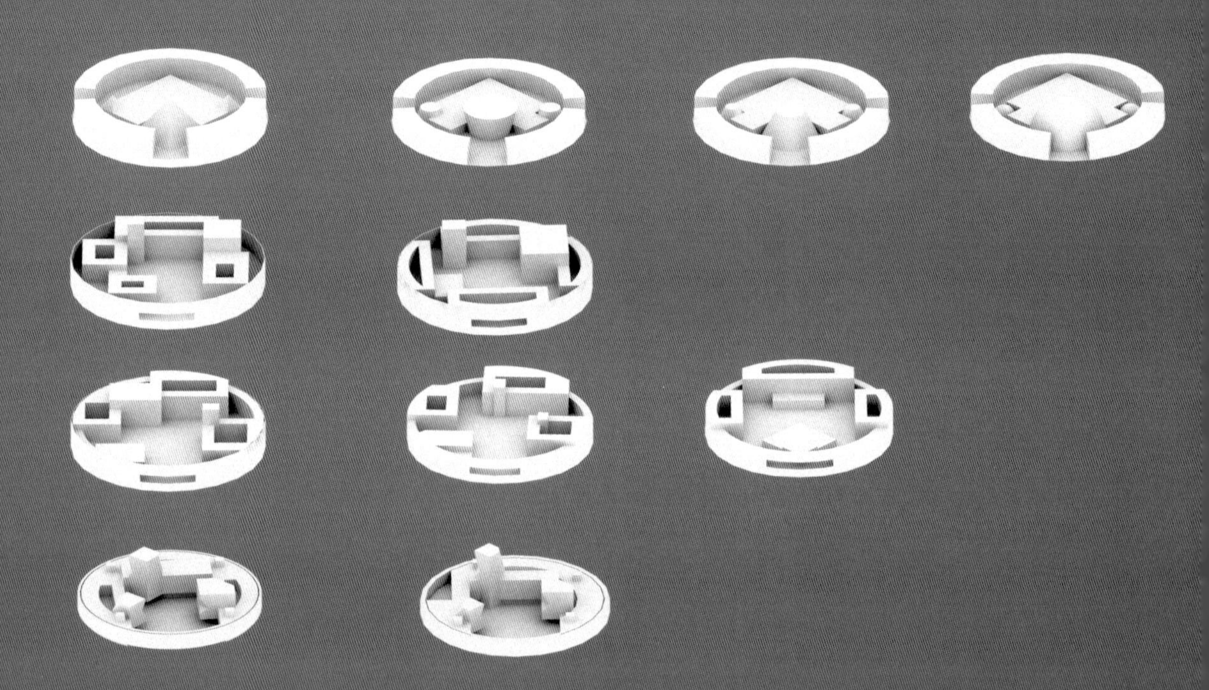

in the winter can provide effective heating, whereas effective systems can also prevent solar radiation in the summer. To achieve an optimal orientation, the design team mapped solar gains and losses for all 360 directional degrees of the site (at an interval of 5 degrees each). Each direction then received one of the following three classifications: receives solar energy in the coldest three months, receives solar energy in the hottest 3 months, and annually average solar energy received. The layout of the design is tended towards balance between winter and summer solar gain. Based on this, not only does the design achieve maximum solar heat in winter, but is also slightly shifted eastward, allowing for protection from solar radiation during summer afternoons.

The Hulun Buir New District is divided into

two upper and lower parts on the Dongshan Plateau. The upper area is classified as grassland with no natural barriers, making wind analysis and wind prevention strategies incredibly important. Directional data is difficult to clearly quantify and display, due to the constantly changing wind direction and speed. A practical solution must therefore take into account wind direction and speed in terms of time and frequency. The design team utilized a wind analysis system to calculate wind from sixteen different directions, using time intervals as units for analysis. Through analysis, it is evident that the strongest winds appear in the springtime but there exists no dominant wind direction. Instead, winds frequently originate from the south and southwest directions. Accordingly, the design team developed a wind proofing strategy which involves a system of windbreaks inside the city limits, as well as arranged urban spatial structure to break the wind organically.

In accordance with the varied conditions on the two upper and lower sites of the plateau, two different mesh systems were selected to generate the basic form of urban space in order to meet differing requirements in terms of solar radiation, wind proofing, as well as other factors. The lower area generally adopted the original existing city grid organization, whereas the upper area adopted a north-south oriented rectangular grid in addition to the continuation of the existing city grid. Not only can both organizational systems of the upper plateau work in conjunction, but the north-south grid also ensures efficient orientation for new designs. In two corners of the grid, staggered systems were set up which achieve two goals: set up wind breaks, and create an ecological environment to stimulate urban vitality. It is still rare to see grasslands areas utilize the practice of shelter-forests. Here, nature and humanity intertwine into a mosaic, constituting the idea of "artificial nature" within the city's spatial frame.

· *The Urban Landscape's Geographical Environmental Simulation*

Urban design does not simply seek beauty on the surface, instead it searches for the intertwining of nature and society as well as coordination and comfort provided by the system of the urban landscape. It is concerned with a great range of urban ecological and artificial systems to coordinate the implementation of a symbiosis between mankind and nature on the basis of a comprehensive environmental management system. Design principles are based upon the natural landscape which then undergoes intervention to achieve the optimization and upgrade of original natural conditions.

The Hulun Buir Plateau contains geographical features typical of a grassland: meadows, steppes, and wetlands. The new urban landscape system attempts to simulate and optimize these features, creating a city with the aesthetic value system of the grassland plateau area.

In the Hulun Buir District, the Dongshan Plateau stands as a majestic background element to the city's east while, to the west, the Yimin River flows from south to north like a ribbon. The Dongshan Plateau and the Yimin River are the city's largest natural landscapes, and play

supporting roles for landscape architecture. In order to strengthen these geographical features, a T-shaped framework was designed around which landscape was organized. First, an island in the middle of the Yimin River was extended eastwards to the ecological green axis on top of the Dongshan Plateau. Then, a waterfront landscape area was established running from north to south along the bank of the Yimin River.

The ecological green axis spans about 600 meters, begins at the Hexin island in the middle of the Yimin River, and continues to the top of the Dongshan Plateau. It differs from normal urban green axes in that it rises gradually, it links the city's two most prominent landscape elements, opened a corridor between the Dongshan Plateau and Yimin River, as well as strengthened and supported originally existing landscape elements. The administrative centre and cultural tourism facilities are organized on either end of the ecological green axis, while the axis itself is raised above ground in order to reduce segmentation or effect on the green axis by the city's roads. This simultaneously provides car parking space below the raised ecological green axis.

A waterfront landscape was created along the bank of the Yimin River. This axis runs from north to south and traverses the travel service, office, commercial, and creative industry areas to intersect with the administrative area forming a plaza. Two border paths preserve the riverfront and its continuous connection with the ecological green belt. Two islands on the river contain facilities for tourism, leisure, recreation, and other functions in other to create an environment which integrates urban life into nature.

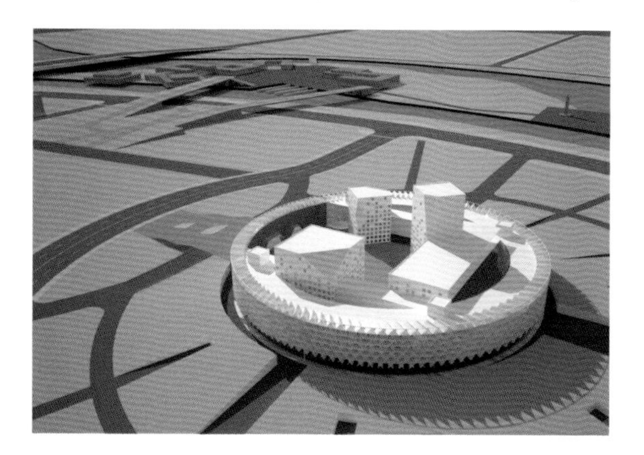

· *An Architectural Language Which Reflects Geographical Features*

Ever since Lewis Mumford proposed Regional Modernism in 1947 as an alternative to the International Style, the idea of regional architecture has been closely observed. Especially due to the globalization and industrialization brought about by the last century, regional architecture has become a common concern. From the perspective of "artificial nature", it can be said that architecture is this "artificial nature" which regulates manmade and natural elements. It not only provides shelter, but also affects spirituality. It is the specific influencer of life and social characteristics in different eras. As a regulator, architecture focuses upon feelings towards and interactions with the environment, the idea of a present feeling, a present scene, or a present place representing architecture's grounding. The grounding of an architecture language into a certain time or place is diverse and comprehensive, and must consider specific geographical and socio-cultural environments associated with ethnicity and local characteristics whilst simultaneously embodying characteristics of the times.

The typical regional architecture of the Hulun Buir region is the yurt, of which the most significant features are the conformation to nature, utilization of nature, and decoration of nature. It not only strongly adapts to local conditions, but moreover utilizes the influence of geographical features upon architectural culture to comprehensively coordinate society and architecture, people and environment, man and nature. Of course, environmental influences are not the only factor affecting architecture, culture plays a large role as well. Mongolian culture contains its own beliefs, wherein an understanding of the universe is based upon the idea of "circular sky circular earth", thereby closely correlating the Mongolian view of the universe, ideological consciousness, aesthetics, and lifestyle. The idea of the "circle" appears everywhere in Mongolian daily life: economic production, religion, and even military formations. The idea of the "circle" has even been deified by the people, it represents a complacency and an embodiment of neither the beginning nor the end, resulting in a sort of timeless beauty. The "circle" represents the Mongolian peoples' yearning and pursuit for wonder, it is the universe, it is the projection of natural mechanisms such as the sun and the moon.

In the design of Hulun Buir, attempts were made to absorb inspiration from Mongolian culture. Architecture was refined to reflect the language of the grassland and prairie life, these concepts were applied to key elements of architecture and urban design. For example, the administrative centre, as an important node which functions as an influencer of the rest of the city, adhered to these principles. Architectural languages were developed with the idea of refined geographical features as well as the unification of outer shape and inner intention to pursue harmony, thereby achieving a wide range of possible architectural identity which produces a sense of belonging and pride. It must be emphasized that this system

防风林景观

Views of the Windbreaks

utilizes geographical features upon a modern language, that it combines the two to demonstrate a new grassland culture.

· *Closing Thoughts*

In order to proliferate an ecological focus in this era, architects need to transcend traditional design thought. New design concepts and ideas rely on the power of analysis, which is dependent on information analysis and the decoding of spatial architecture and natural forms. The evolution of urban and architectural spaces requires logical and precise deduction, allowing for space to be defined by external social, economic, and administrative factors. Dean of the Princeton School of Architecture Stan Allen proposes that design is an artificial ecology, wherein the design of activities is akin to resource processing, to the complex interactions between the species and the climate. Architecture is not simply an independent man-made artifact, but rather the blurred boundary between architecture and landscape. In terms of cities as an "artificial nature", planning and design need to explore opportunities provided by the environment, and break out of the mold created by structure and infrastructure. The blurring of boundaries between architecture and landscape as well as city and landscape must be blurred. From here can be sought innovative solutions which integrate the city and the environment, methods which allow for the design of practical eco-cities.

Originally published in *World Architecture*, 2010 (09), with the second author Perry Yang

望京电子城 B12 和 C3 地块
B12 & C3 Plots of Wangjing Electronic City

主题：绿谷

Theme: Green Valley

B12 和 C3 地块是北京望京电子城向北扩展的区域，其位于北五环路与东五环路的连接处，是两条道路的视线交汇点，对城市景观至关重要。

本方案依据“绿谷”的理念来整合两个地块的空间结构。若干栋规模不等的办公建筑分散设置，以满足不同规模企业的要求；为企业提供公共服务的部门，如会议、餐饮、管理服务等，集中设置，以方便交流、共享资源。公共服务部门设置于“绿谷”之下，“绿谷”将各栋独立建筑连接起来。“绿

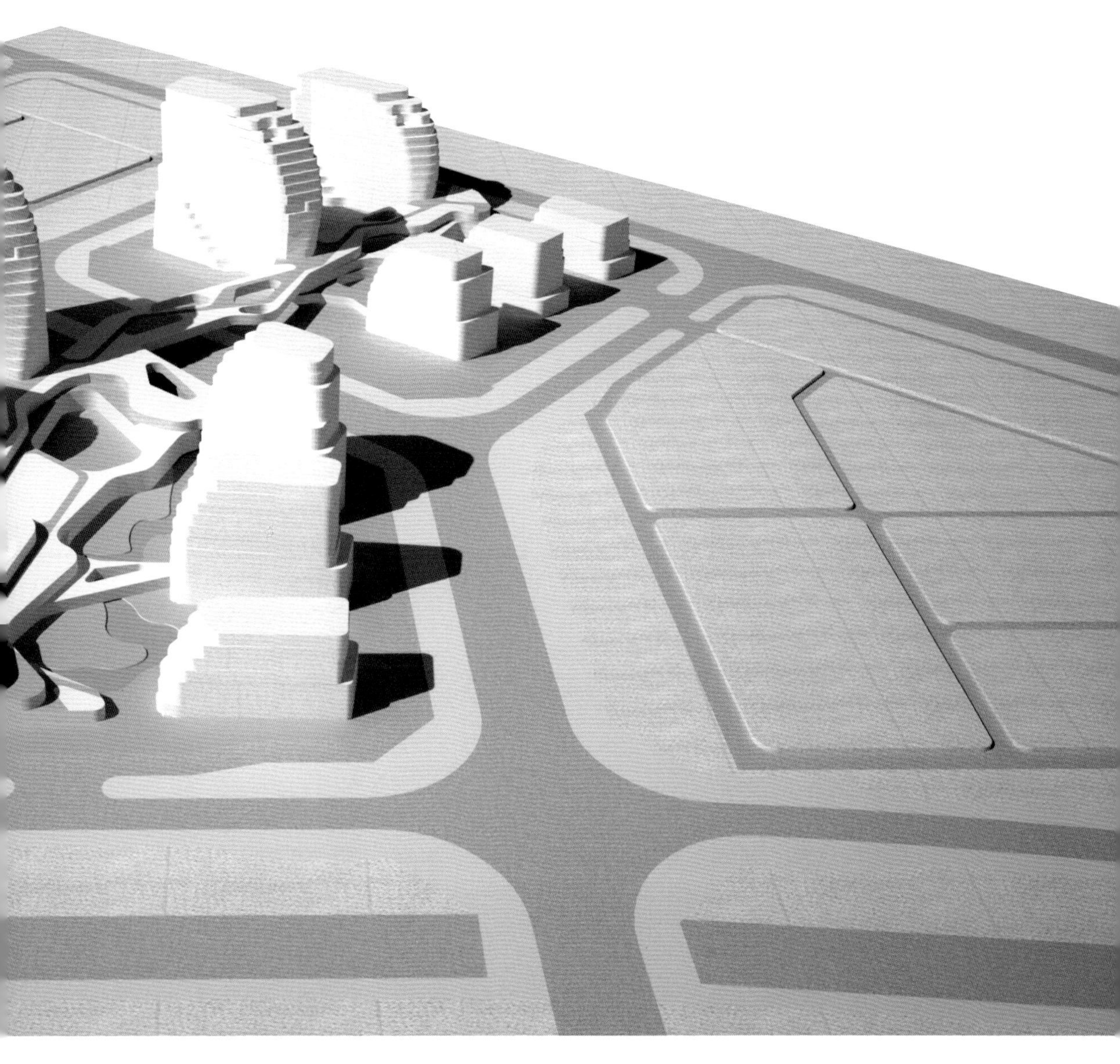

谷”将使整个园区景观质量得到很大提升。

The B12 and C3 Plots are the north exparsion of Wangjing Electronic City; They are located near the junction of the North Fifth Ring Road and the East Fifth Ring Road. The B12 & C3 Plots are very important to city landscaping, as they are also the visual junction of two Roads.

The project relies on the "green valley" concept to integrate the spacing structure of the two plots. A few commercial buildings of different sizes scatter to fulfill the needs of corporations with different sizes; the project provides public service departments to corporations such as conferencing, catering, and managing. The pubilc spice is in the centre for easy communication and shared resources under

立面图
Facades

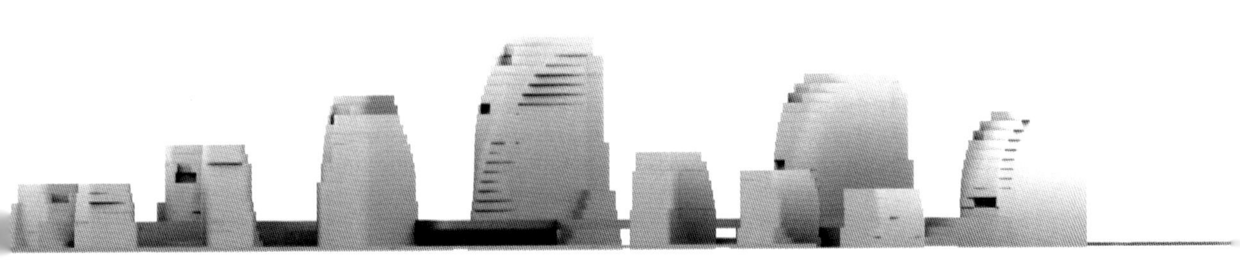

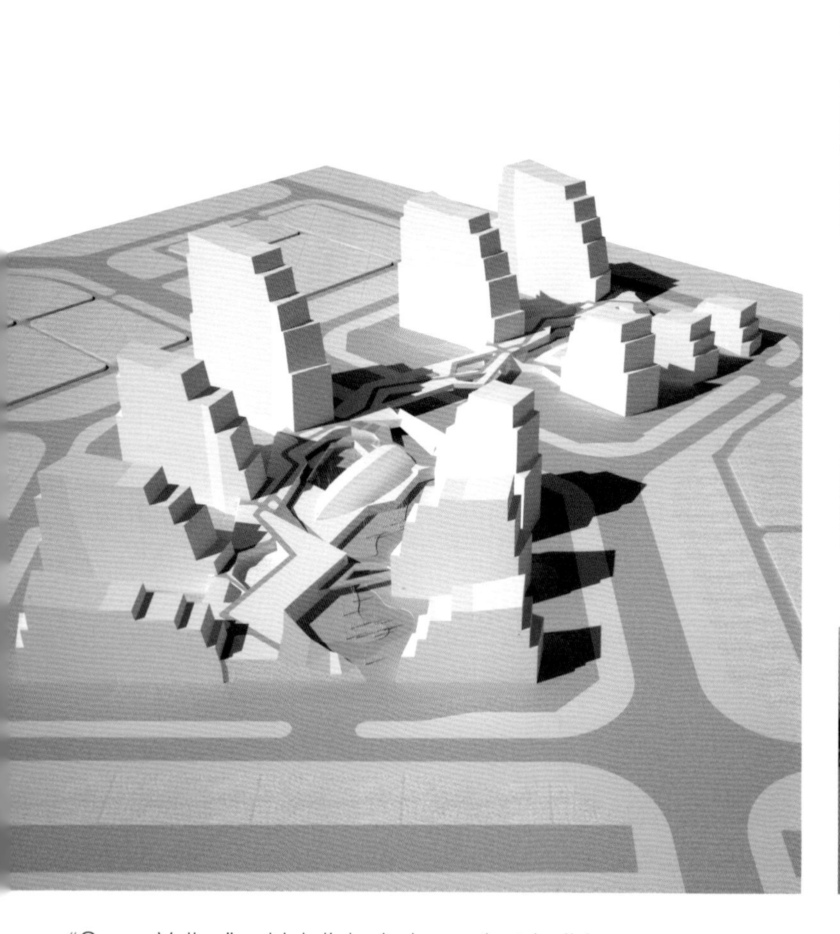

"Green Valley", which links independent buildings. Project "Green Valley" will promote the quality of the landscape park by a large margin.

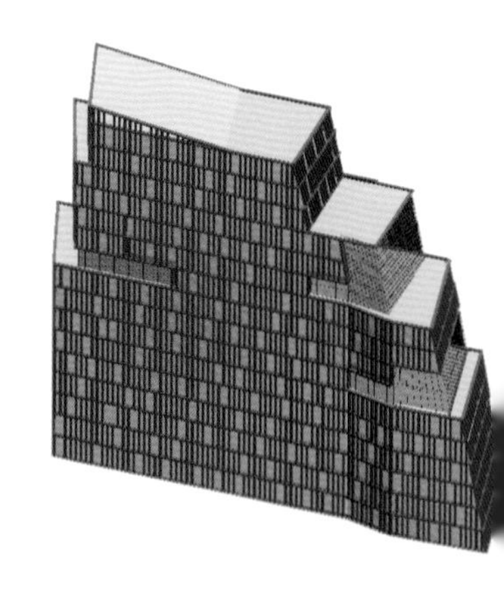

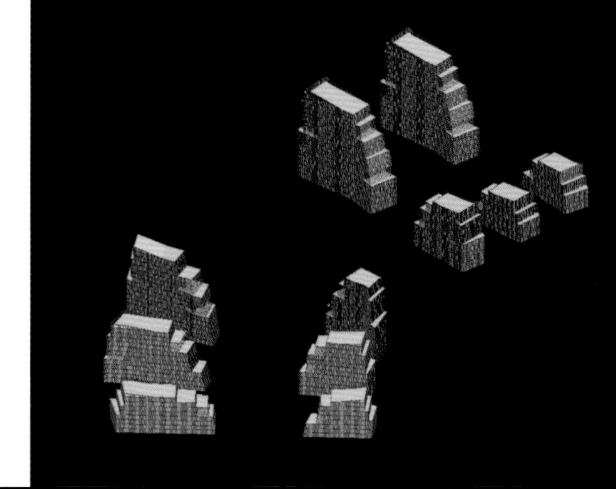

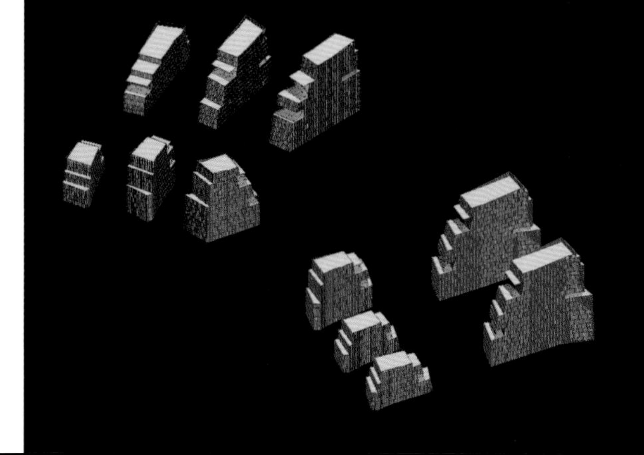

本土地景 LOCAL LANDMARK
文脉更新 CONTEXT RECONSTRUCTED
本土建构 LOCAL TECTONICS
城市空间 URBAN SPACE
本土再造 LOCAL UPCYCLING

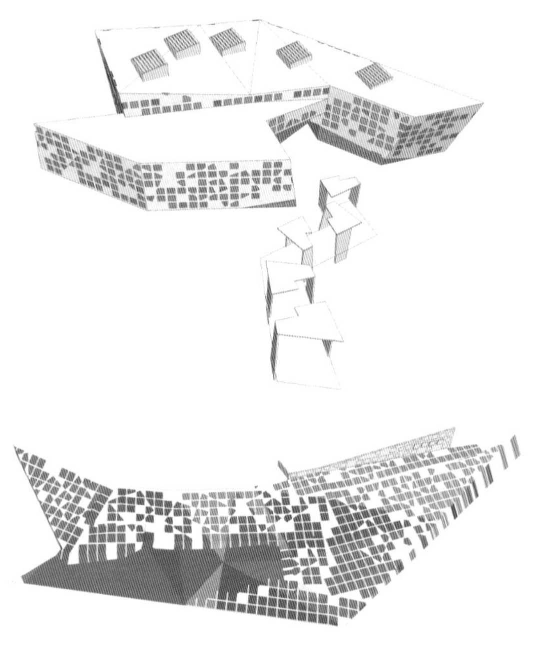

大城小院会所

Dacheng Xiaoyuan Club

主题：一个景观焦点

Theme: A Landscape Attraction

大城小院会所位于柳州市桂柳路与学院路交叉口，处于城市景观轴线的延长线上，是城市重要的景观节点之一。本方案力图通过使用当代语言，对基地进行诠释，使建筑成为一个引人注目的城市景观节点。

一个形如巨石的建筑体量居于基地之上，两片折墙将巨石劈开。建筑被分为若干“形体”，“形体”之间进行“对话”。本方案充分利用了建设基地的地形高差变化，让地形和建筑形体之间穿插呼

应。设计赋予“形体”自由鲜明的个性特征，但又将所有“形体”统一于基地的脉络之中。建筑外表皮图案是通过简单图形的变异形成，还有个性特点。

Dacheng Xiaoyuan Club is located next to the intersection of Guiliu Road and Xueyuan Road in Liuzhou City. The club rides on the extension of the city axis, acting as a major landscape attraction. The project seeks to interpret the base with design language of the current era, making the architecture an eye-catching landscape of the city.

A building mass sits on the base like a giant rock, splitting two accordion walls. The architecture is divided into several “shapes”, and “conversation” is held between “shapes”. The project makes full use of the height difference in the foundation of the architecture, produc-

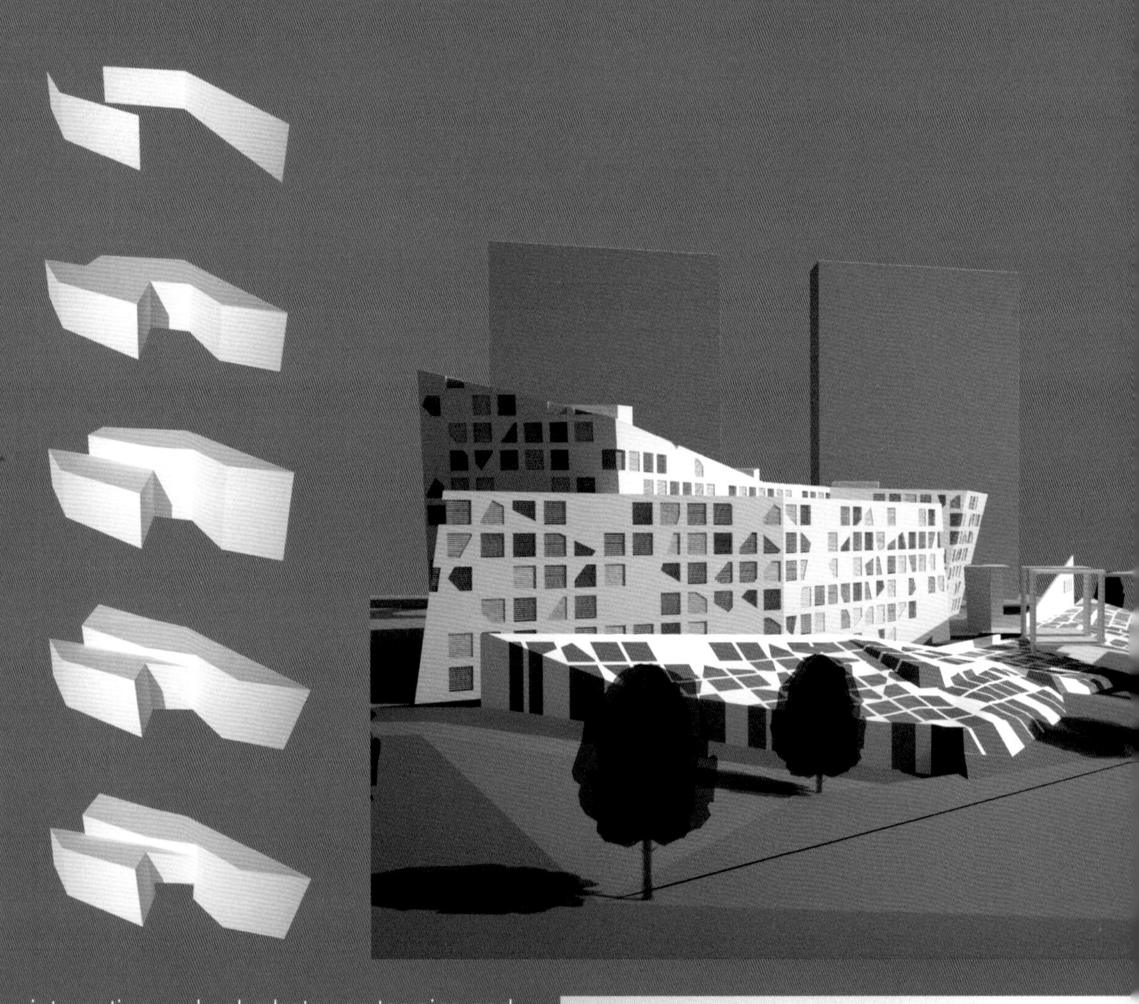

ing interaction and echo between terrains and buildings. The "shapes" are given distinctive yet free personalities by design; however, all "shapes" are united in the skeleton of the foundation. Patterns on the building exterior are varied from simple figures to show stylish characteristics.

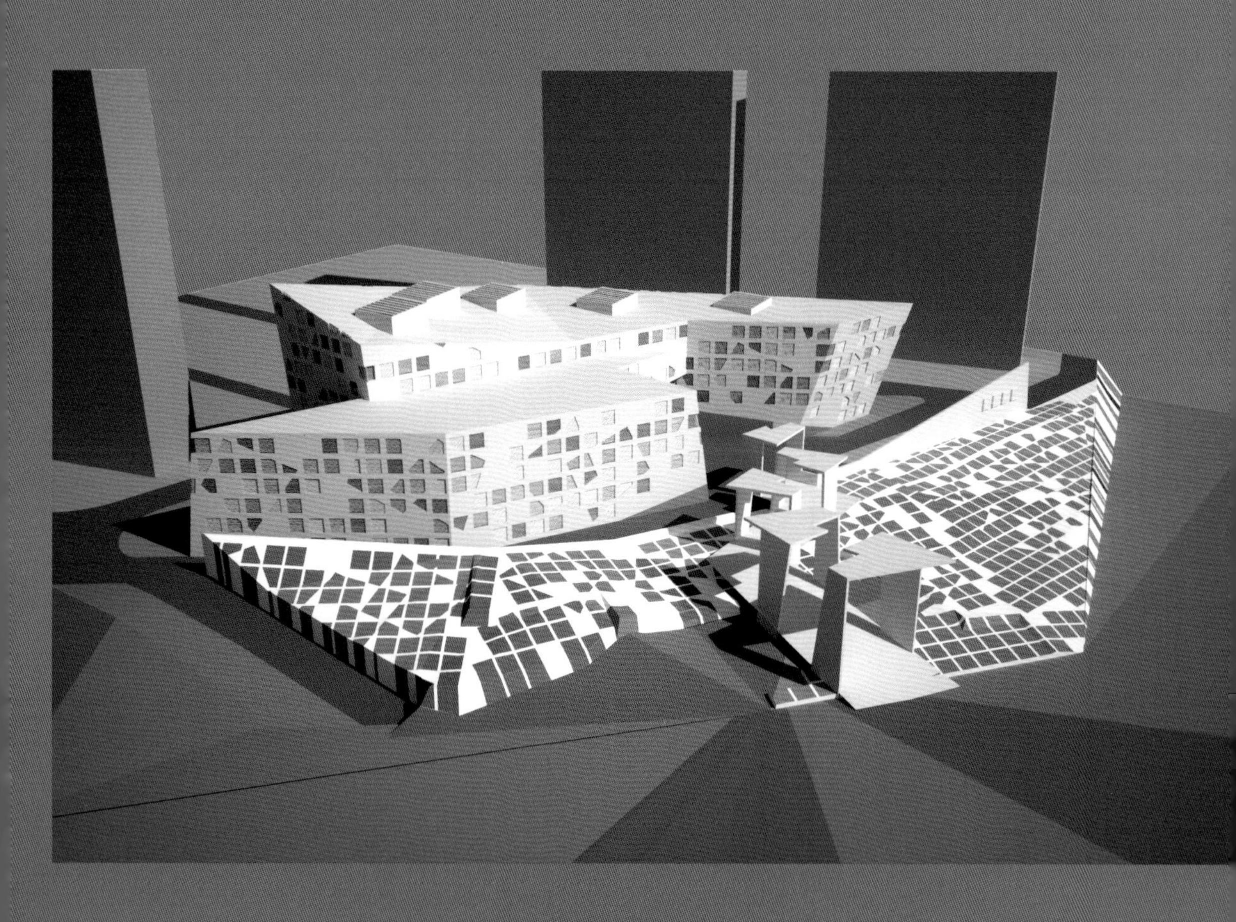

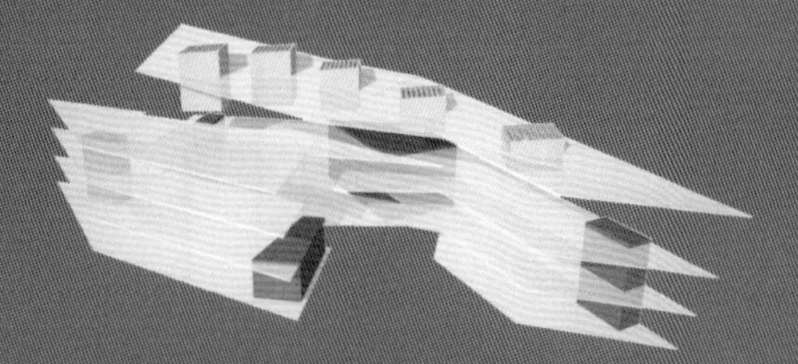

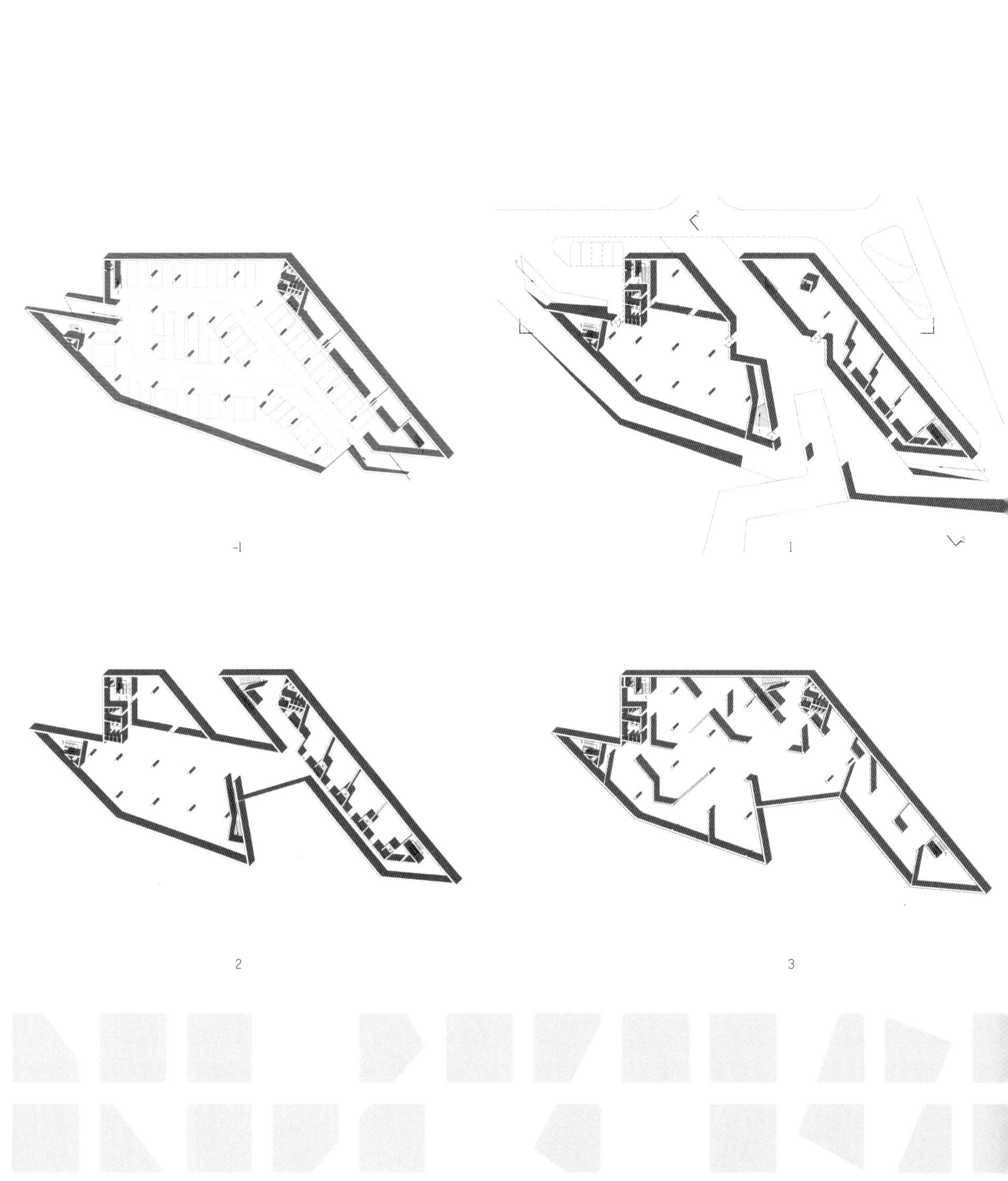

-1
1
2
3

立面图 & 剖面图
Facades & Sections

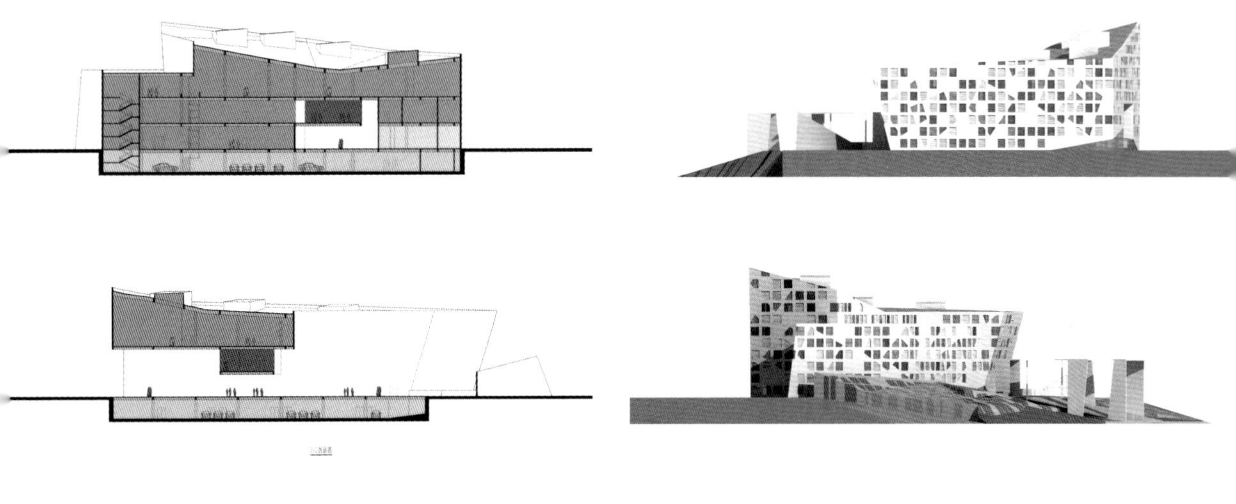

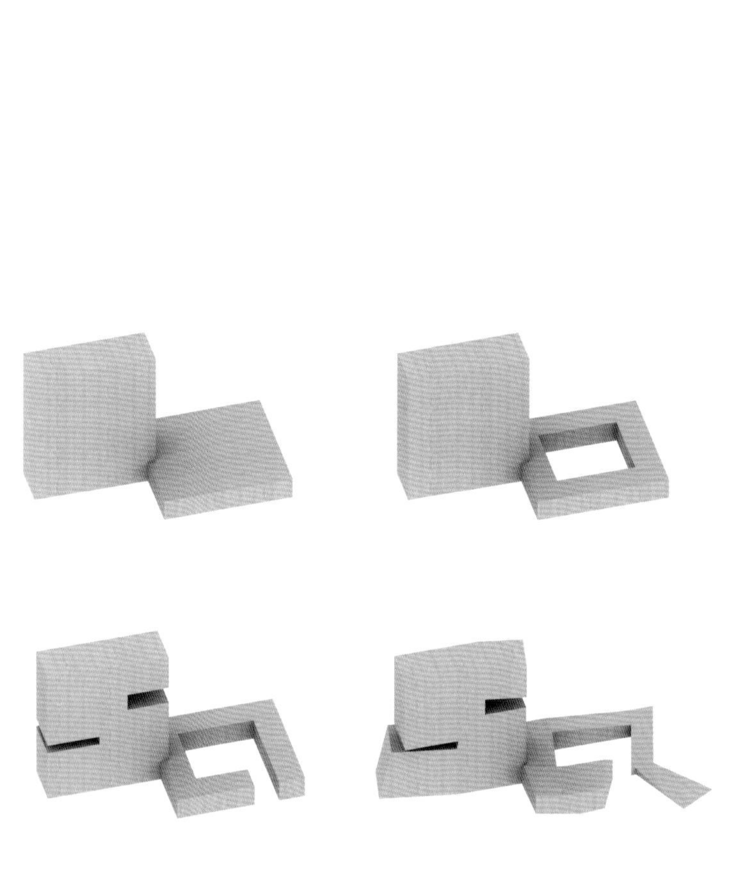

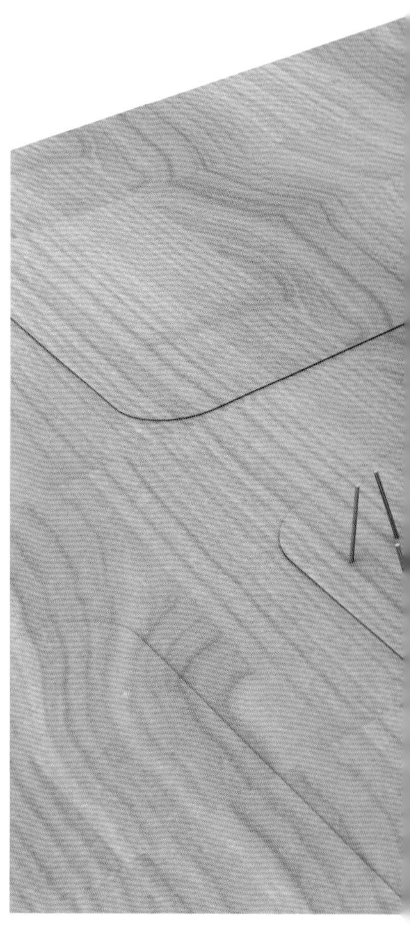

唐山市质监综合楼

Complex of Tangshan Quality Supervision

主题：Q+S

Theme: Q+S

该项目位于唐山市凤凰新城。在建的凤凰新城一片空白，棋盘式的道路没有多少可借鉴的文脉。如果单体建筑把握不好，将形成火柴盒林立的城市形象。国际经验证明，板式办公楼较之于塔式办公楼在节能方面更具优势。为了避免呆板的火柴盒建筑，本方案对板式办公楼做适当变形，“S”状的板楼和“Q”状裙房组成具有张力的造型，并巧妙地将 Quality Supervision 的意义融入其中。另外，建筑外表皮南北向采用不同的肌理，减少能源流失。

The project is located in New Fenghuang Town, Tangshan City. The New Fenghuang Town is a blank canvas under construction. There is no context available as reference for the chessboard style roads. The city image will easily become an array of matchboxes if independent buildings are not handled appropriately. Prior international experiences prove that slab-type office buildings are more efficient on energy conservation than tower-type ones. In order to avoid a dull matchbox, appropriate modifications were made on this design. The S shape slab and Q shape skirt provide the contrast in appearance and meanwhile brilliantly fuse the meaning of Quality and Supervision into the building. In addition, different textures were used in the north and south exterior to utilize sunlight while reducing energy lost.

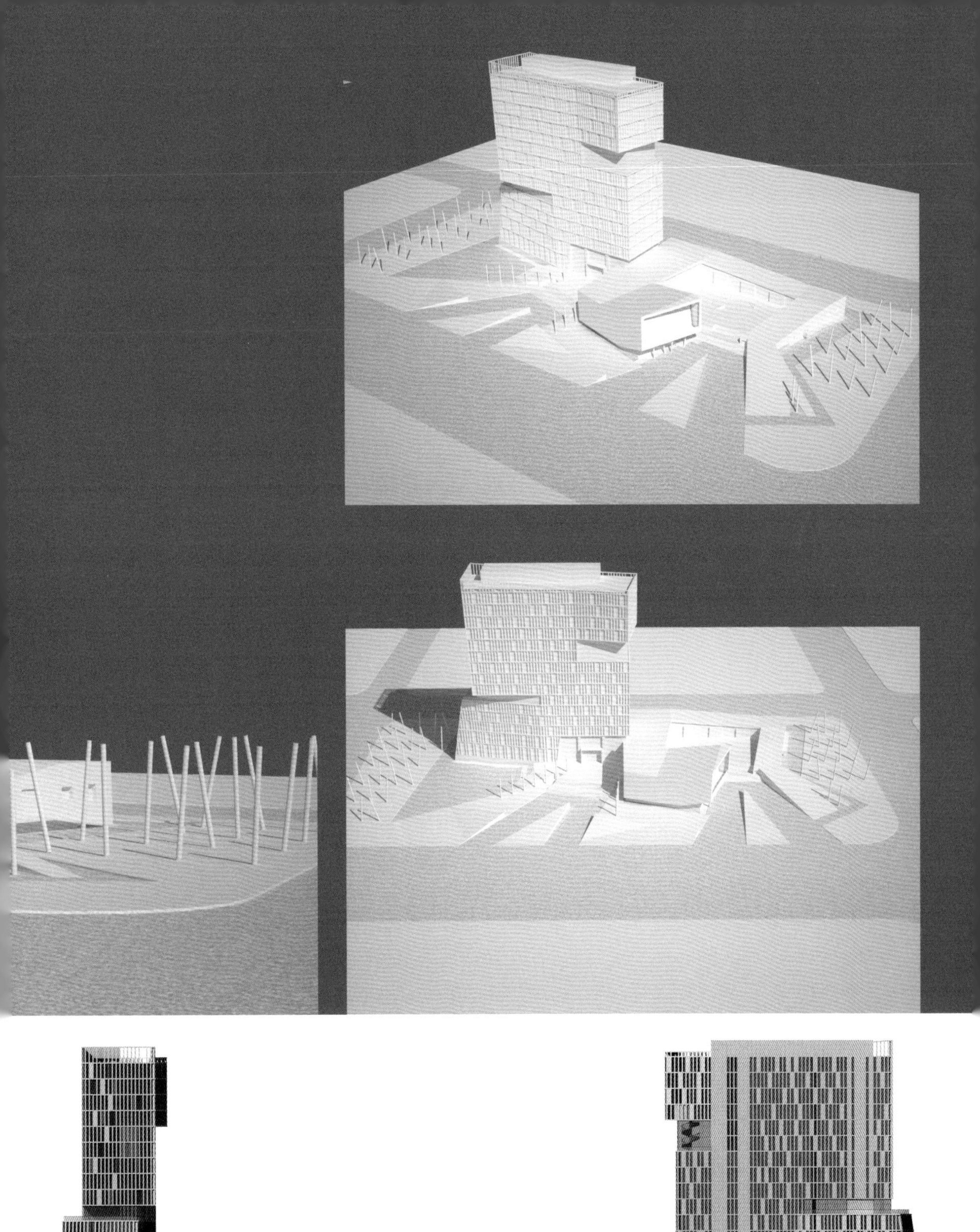

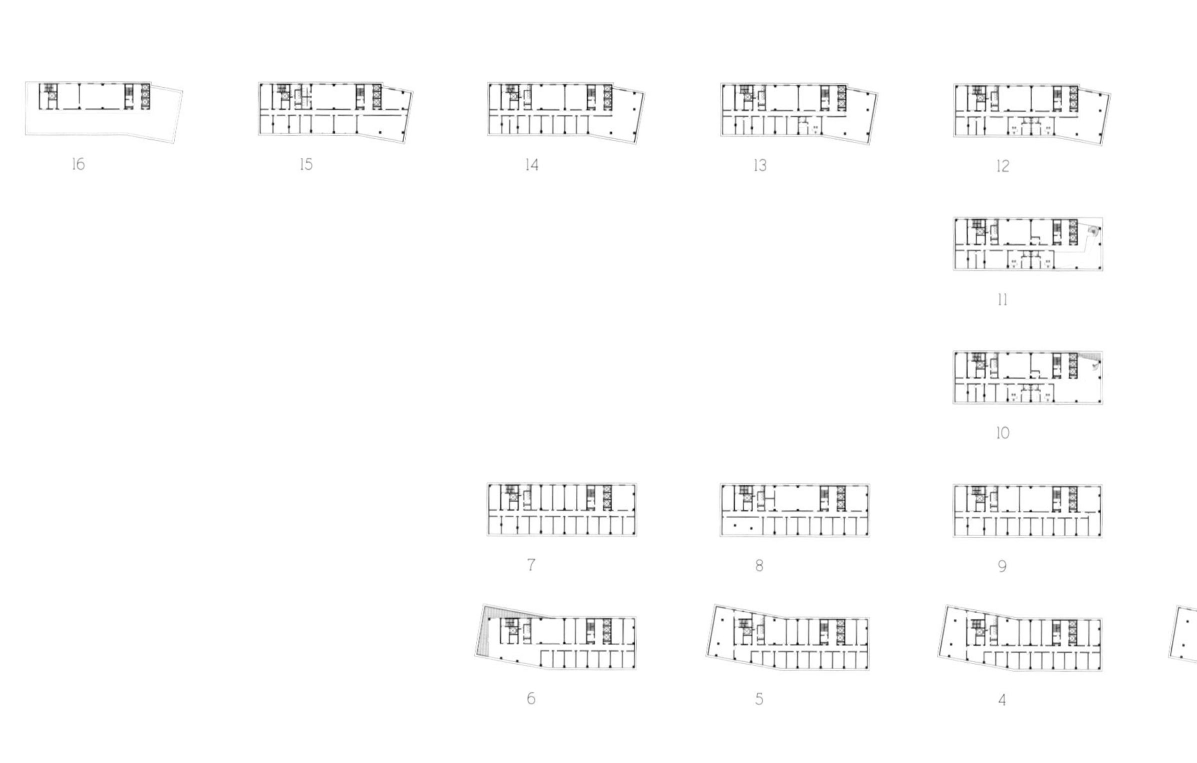

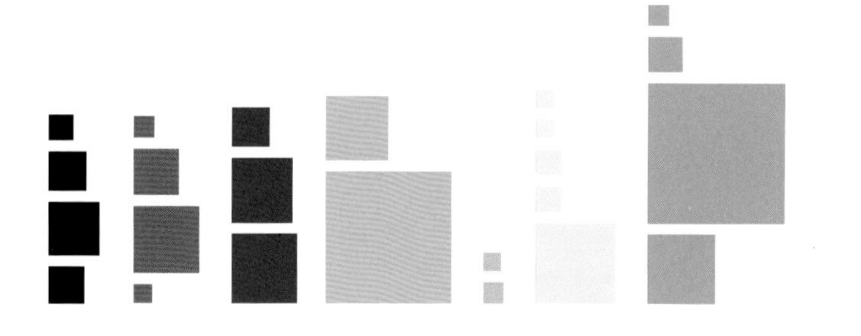

功能组合
Functional groups

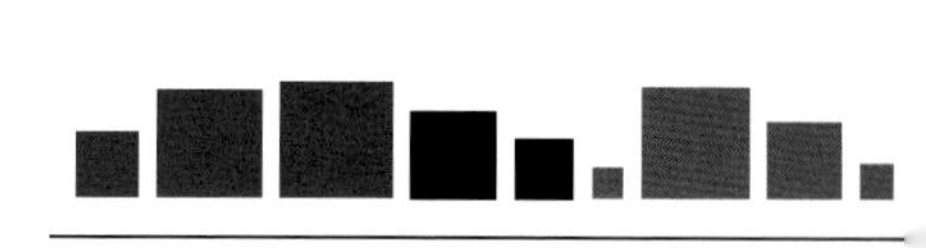

剖面功能组合
Functional groups in section

线性功能组合
Linear functional groups

剖面图
Section

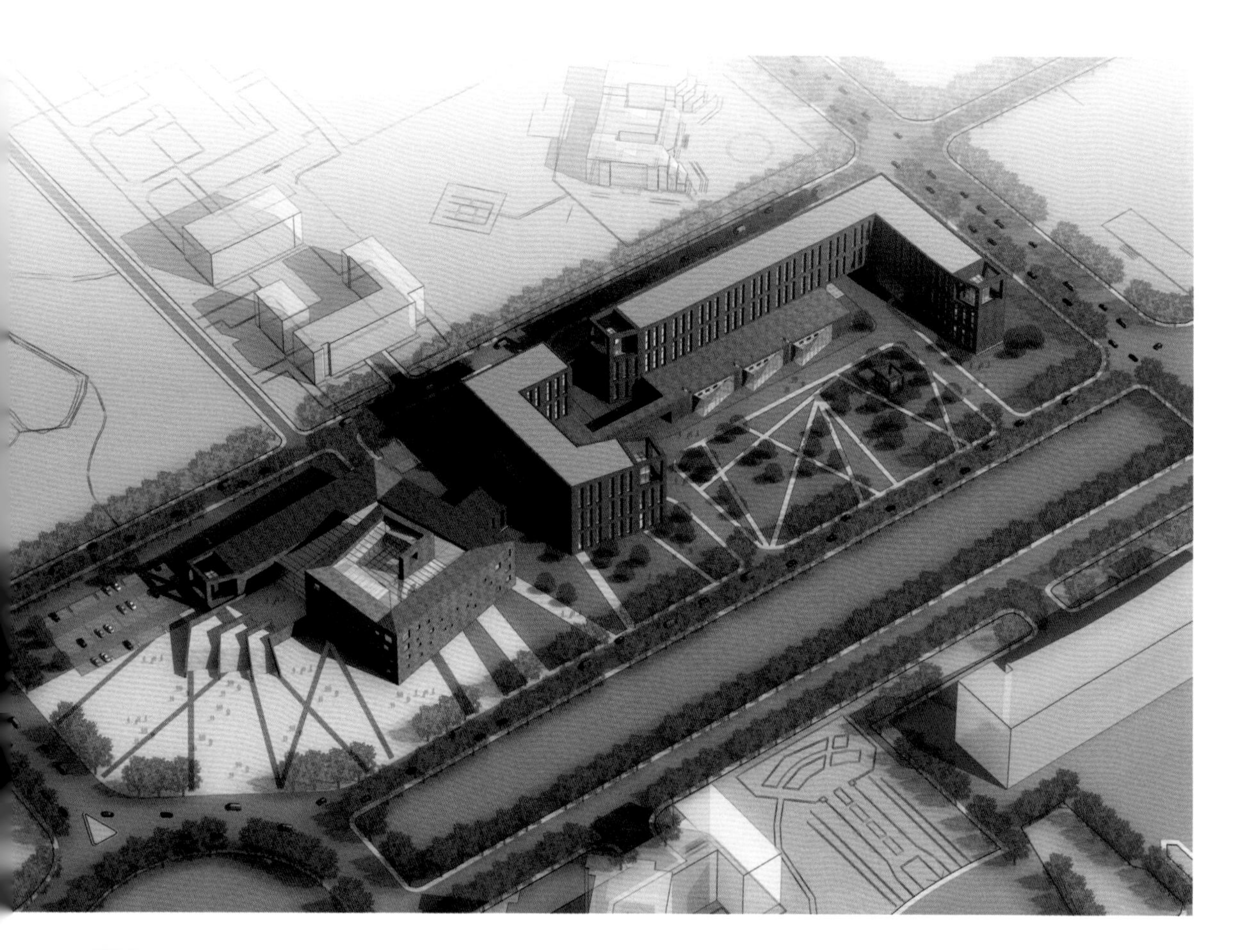

大庆市规划展览馆
Planning Exhibition Hall of Daqing

主题：围合
Theme: Enclosing

大庆市规划展示馆位于大庆市高新区，场地空旷，建筑稀少。大庆市冬季寒冷漫长，对于大庆的地域条件来说，绿色环保的概念是保温和防风，以及改善场地条件，利于植物生长。

大庆市规划展览馆附属有市档案馆、市城建档案馆、规划局等办公建筑。规划展览馆单独布置在地段西侧，面向环岛，利于交通，造型围合，利于保温。其余几栋办公楼连为一体，在其南面围合出日照良好、防风避寒的活动场地和绿化用地。

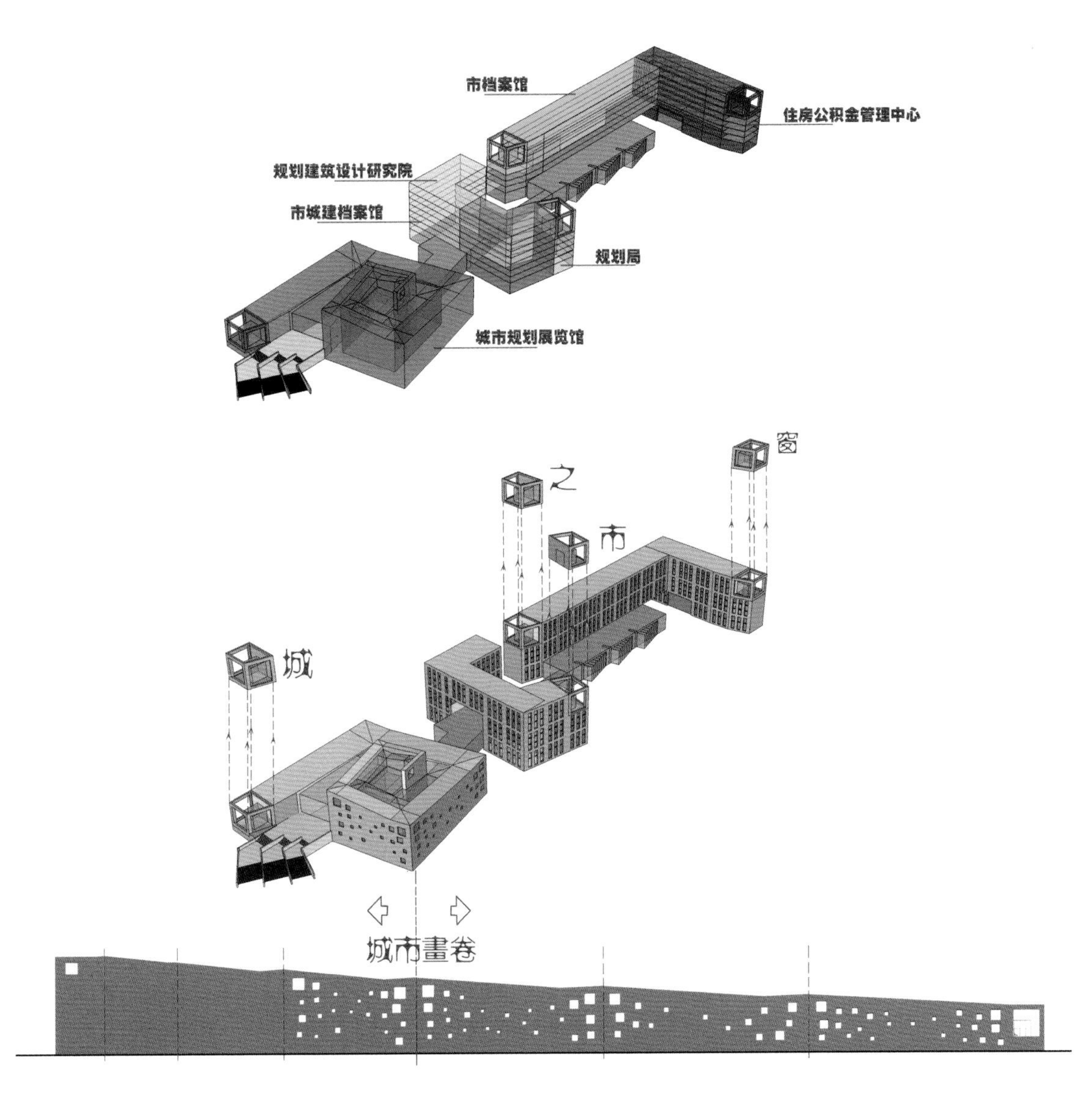
市档案馆
住房公积金管理中心
规划建筑设计研究院
市城建档案馆
规划局
城市规划展览馆
窗
之
市
城
城市畫卷

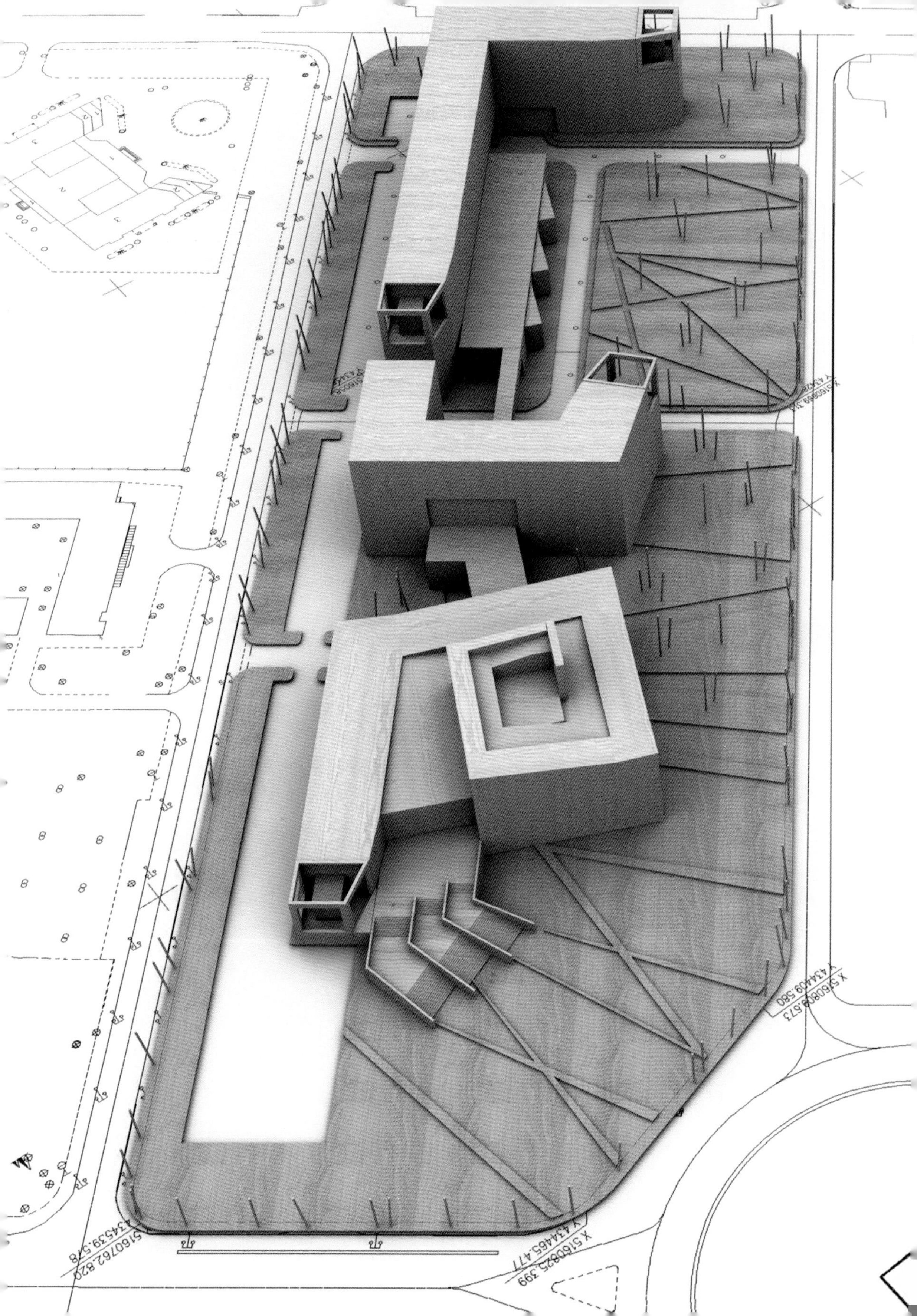

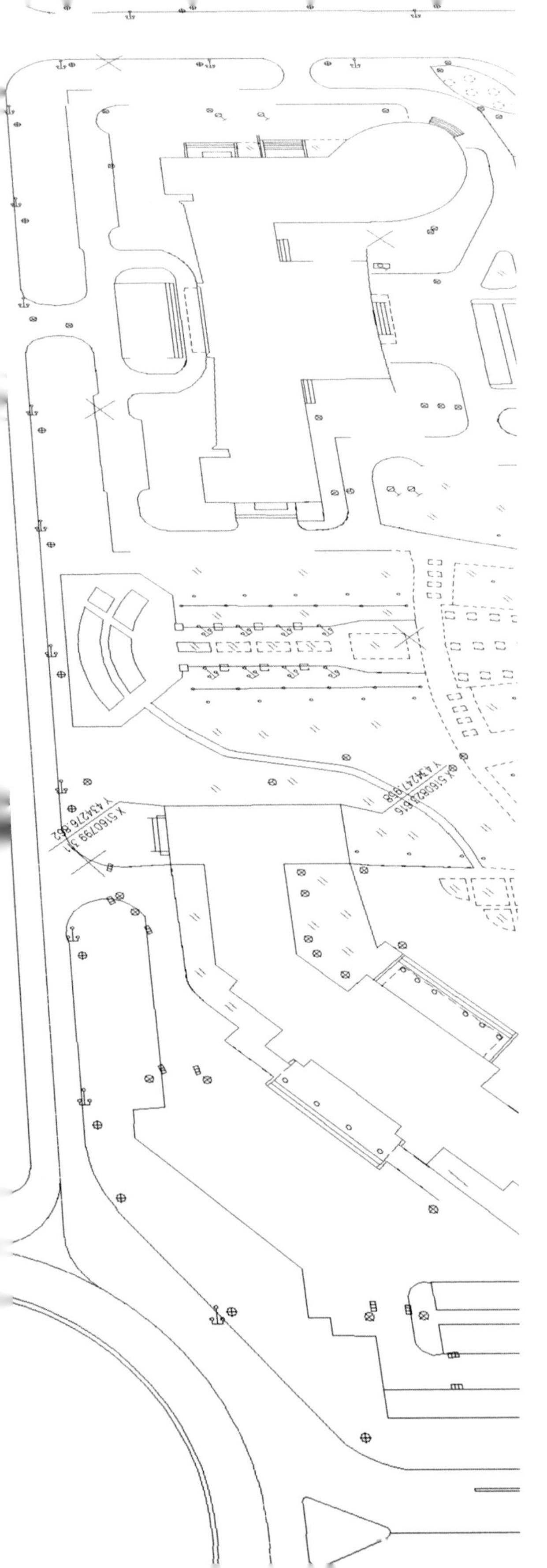

The Planning Exhibition Hall of Daqing is located in an open area with minimal architectures inside Gaoxin Town, Daqing City. Winter in Daqing is cold and long. Considering its local condition, the ecological concept would be heat preservation and wind resistant; In addition, improving the soil to benefit plant growing is essential.

The Planning Exhibition Hall of Daqing pos-

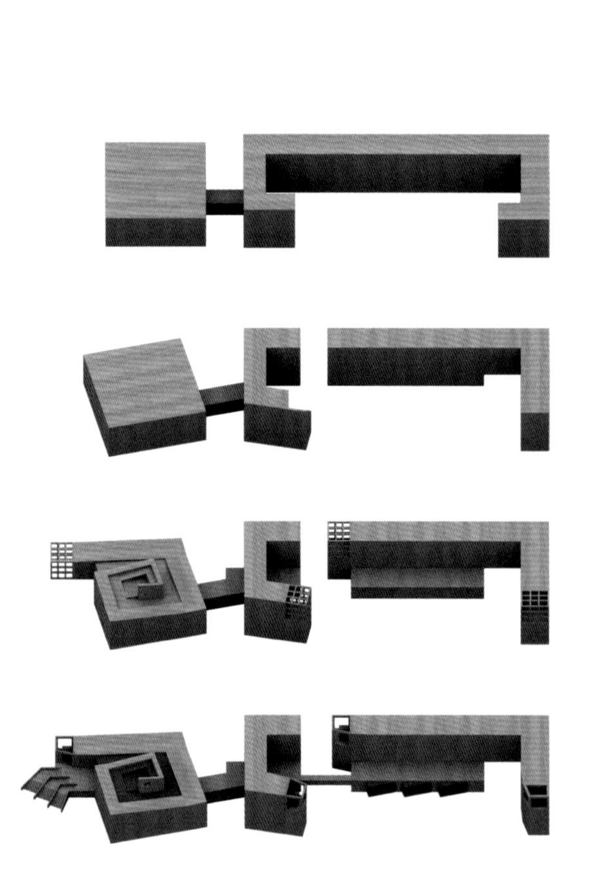

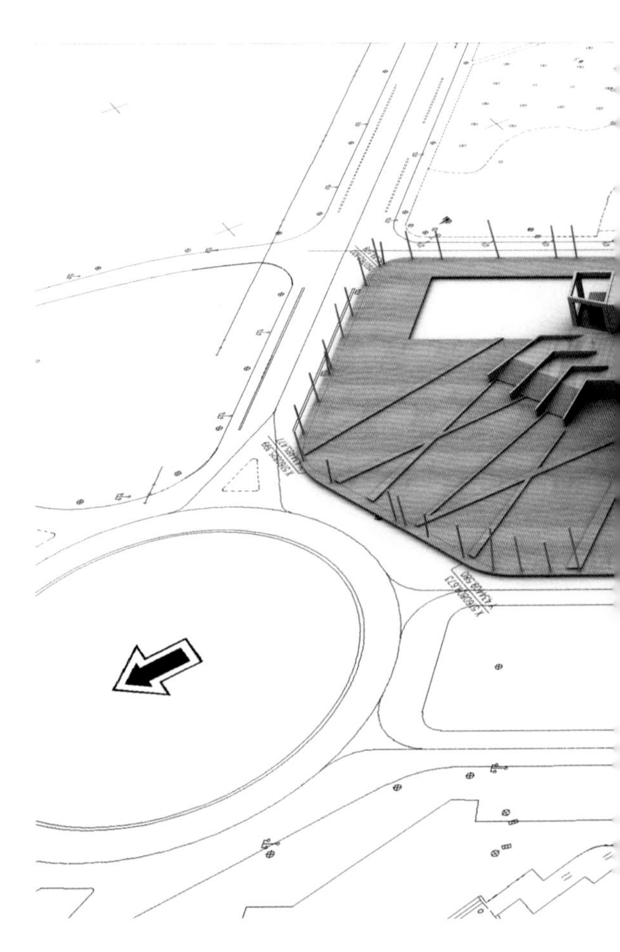

sesses commercial buildings such as City Archives building, Urban Construction Archives and City Planning Board. The Planning Exhibition Hall sits by itself on the west side of the area, facing the island for convenient commuting; its profile is enclosing for preserving heat. The rest of the commercial buildings outline an enclosing arrangement, creating a rich-in-sunlight, and wind and cold resistant public green zone.

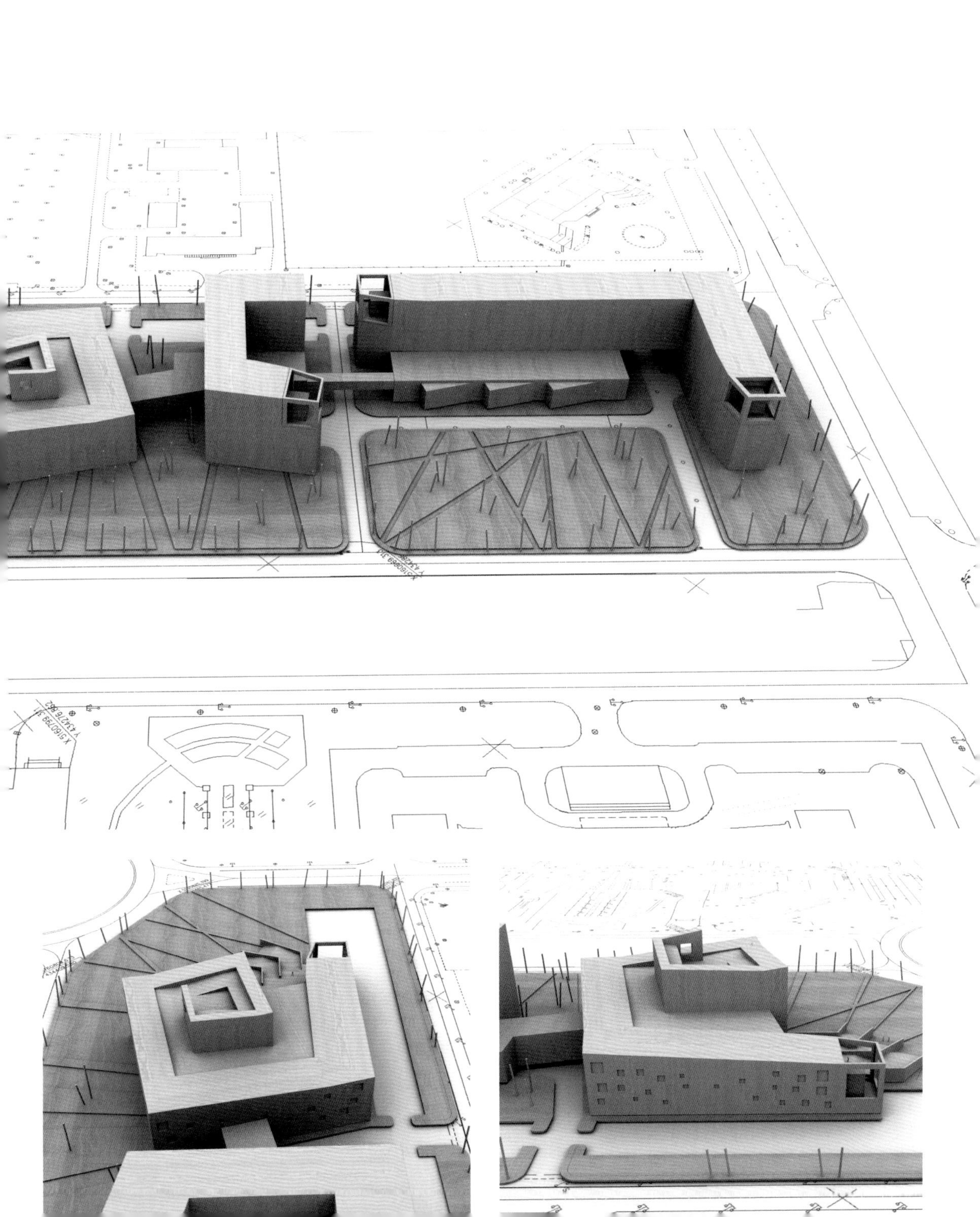

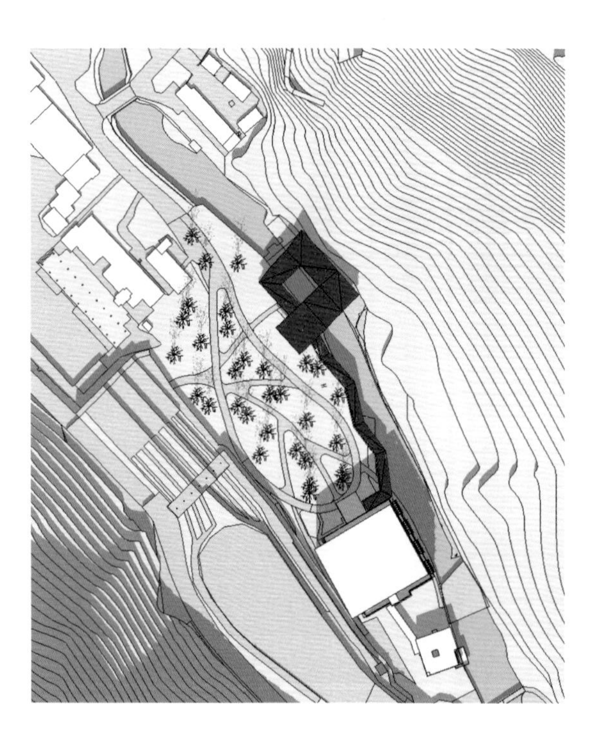

九寨沟贵宾楼水上餐厅

Water Restaurant of Jiuzhaigou Grand Hotel

主题：不与自然竞争

Theme: No Rivalling the Enviroment

水上餐厅靠近九寨沟风景区入口，是现存九寨沟贵宾楼的配套加建项目。在风景区入口这样敏感的地段修建建筑，尊重和理解周边环境，不与环境竞争，是本设计遵循的准则。

建筑使用轻型钢结构建造，并在高度上与水面保持一定距离，以免对河岸造成过多影响。建筑造型与山体呼应，每一个视角都是一幅画面。我们非常注重地段的拓扑特征，所有策略的目的是让建筑以对环境最小的影响融入自然。

The water restaurant is a new addition to the Jiuzhaigou VIP building near the entrance of Jiuzhaigou Touring Area. When building on such sensitive area such as entrances, the project follow the rules of comprehending, respecting and not rivalling the environment.

The project is built with a lightweight steel structure. Height of the structure has been adjusted to keep safe distance to the water

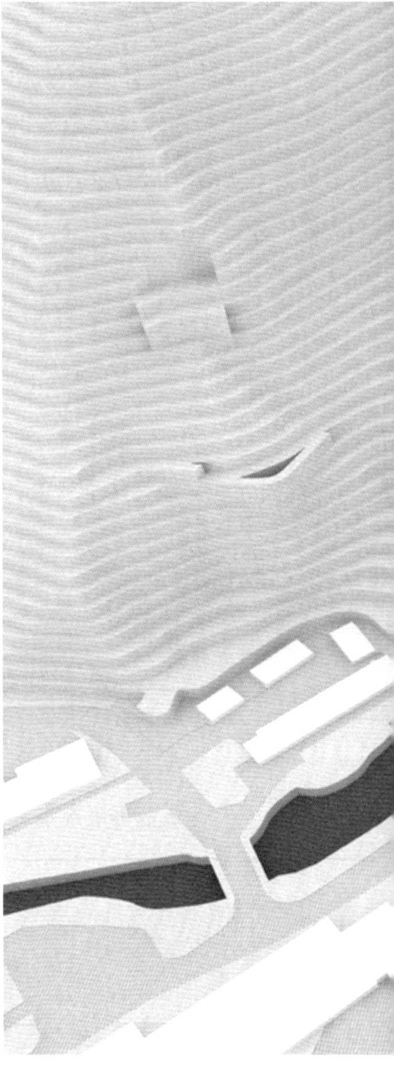

surface, avoiding unnecessary impact to the water body. The shaping of the building echoes with the mountain, creating a different picture from every angle. We focus greatly on the topological features of the site, all policies aim to enter the nature with minimal environmental impact.

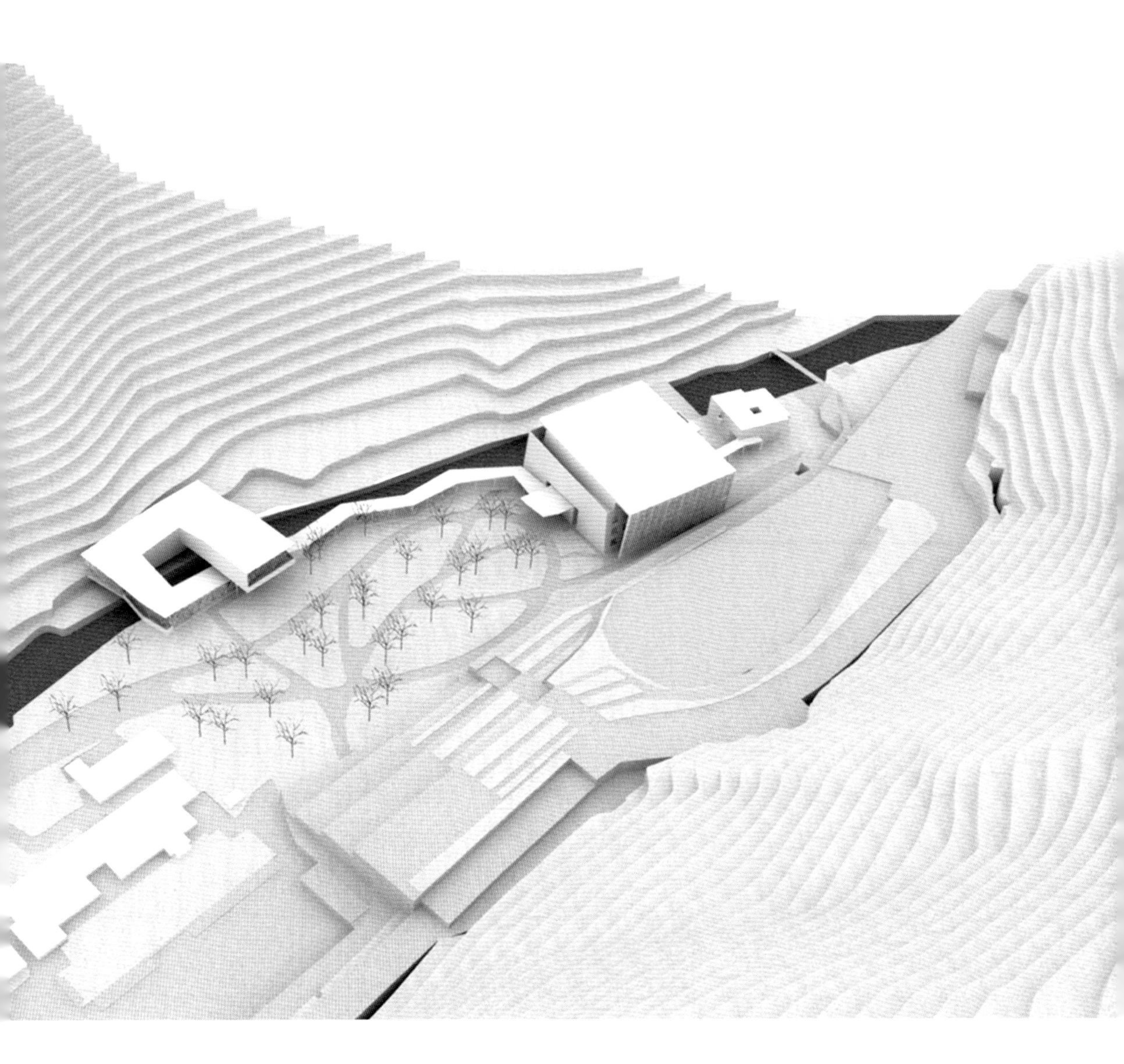

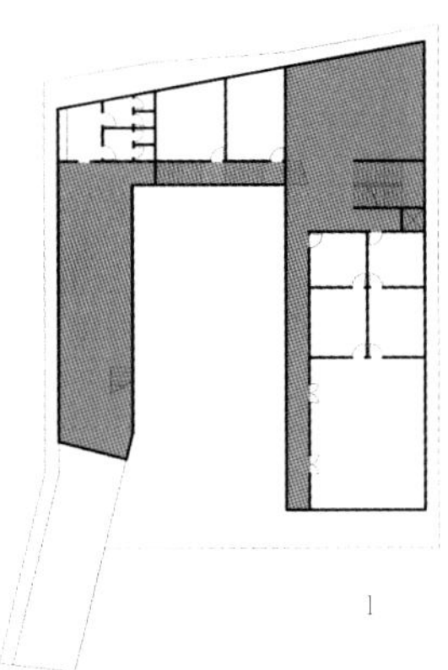

1

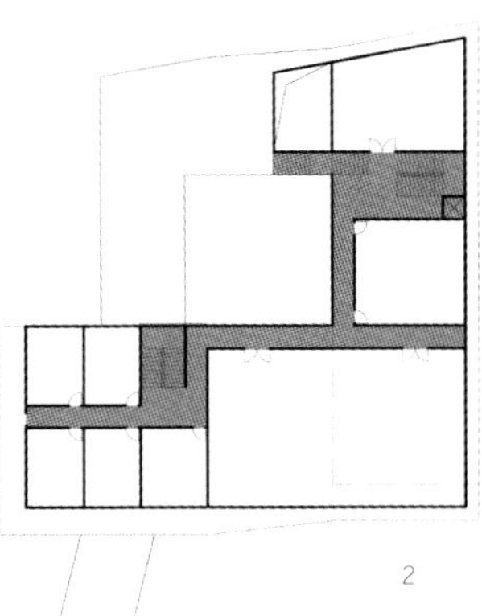

2

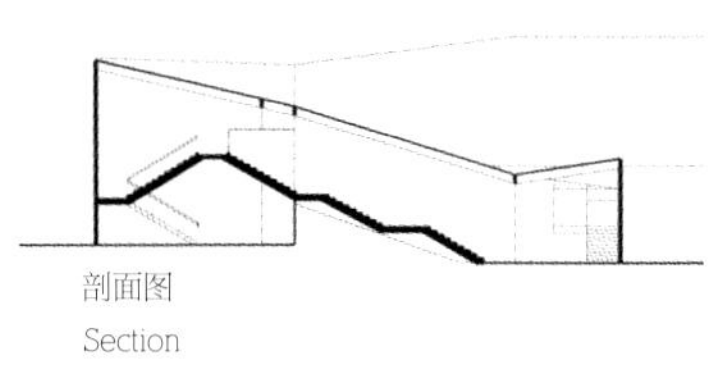

剖面图
Section

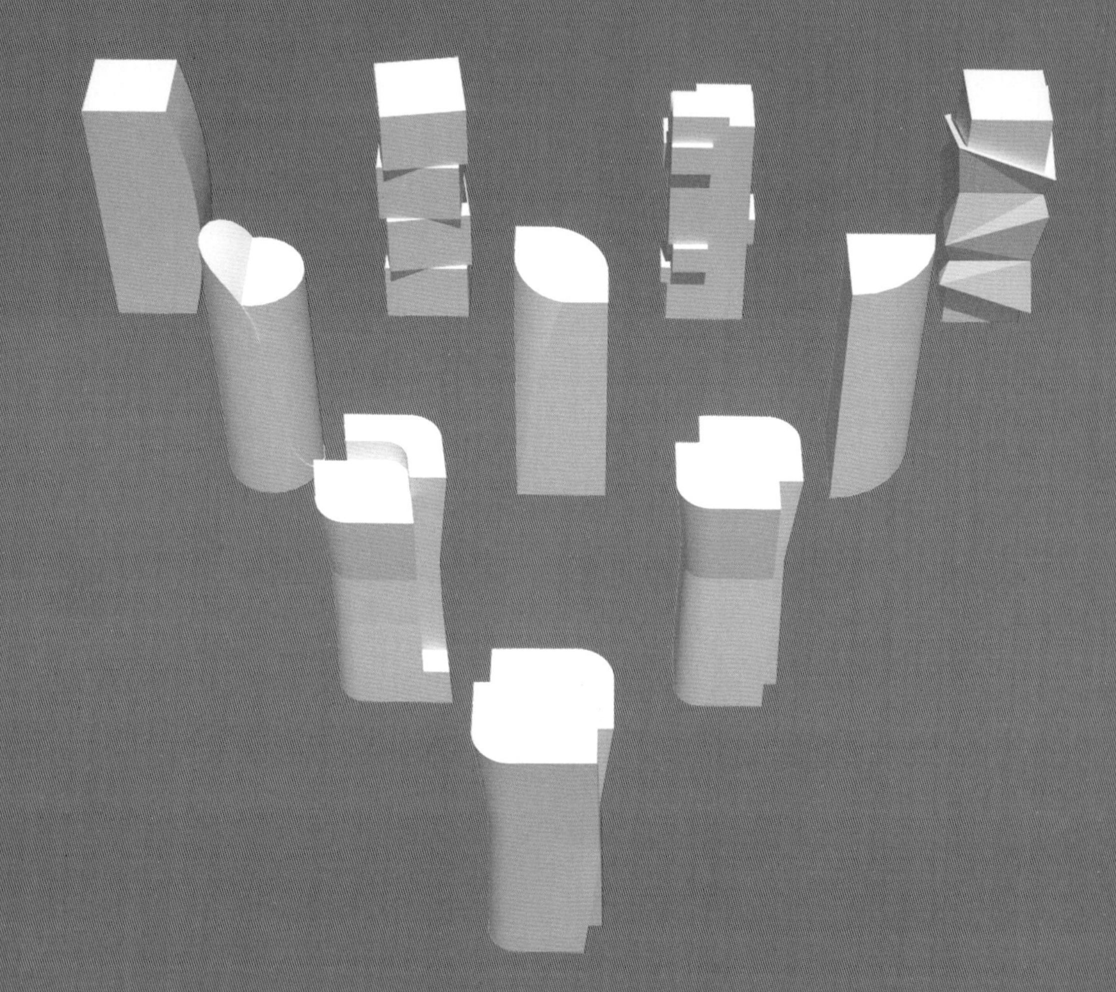

东方博特酒店（五星级）
ORIENTAL BOTE HOTEL (5-STAR)

主题：地区新形象

Theme: A New Image of the Area

进入21世纪以来，北京市产业结构进行了调整，一些污染严重、能耗巨大的企业逐渐搬出市区。首都钢铁公司的搬迁，以及采矿业的关停，使北京西部的城市环境得到很大改善。随着长安街向西的延伸、莲石路和阜石路两条城市快速路的开通，门头沟区与北京城区的联系愈加紧密。

建设中的东方博特酒店靠近长安街西延线，位于石门营环岛东北侧，是从北京城区进入头沟区的门户。为了展示门头沟区的新机遇、新发展，改变

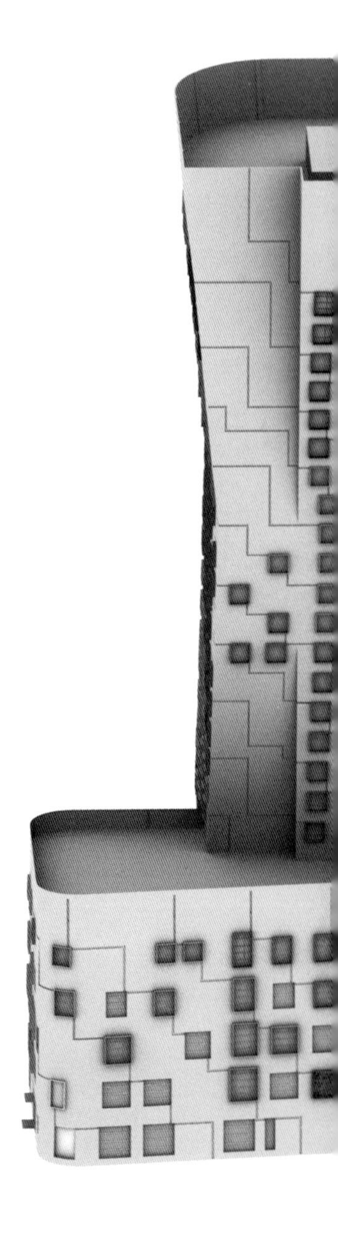

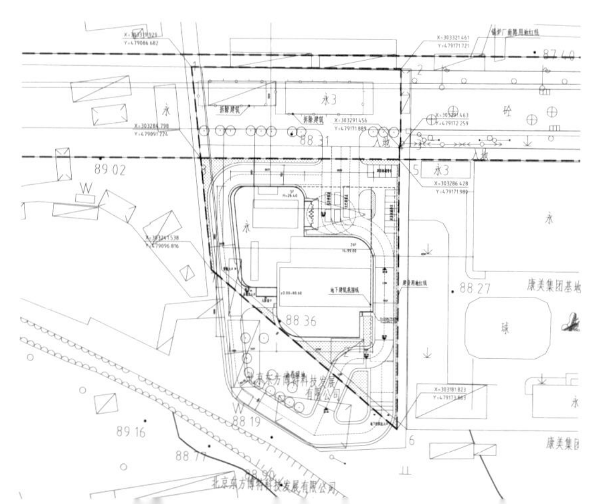

以往人们对门头沟区环境污染、发展落后的印象，本方案采用现代时尚的设计手法，着力塑造个性新颖的建筑形象。但这个形象并不完全是凭空捏造，它与该区域的自然山水以及原有的钢铁工业气质相吻合。

From the beginning of 21st century, industrial structure in Beijing City has been heavily revised. Several enterprises with heavy daily

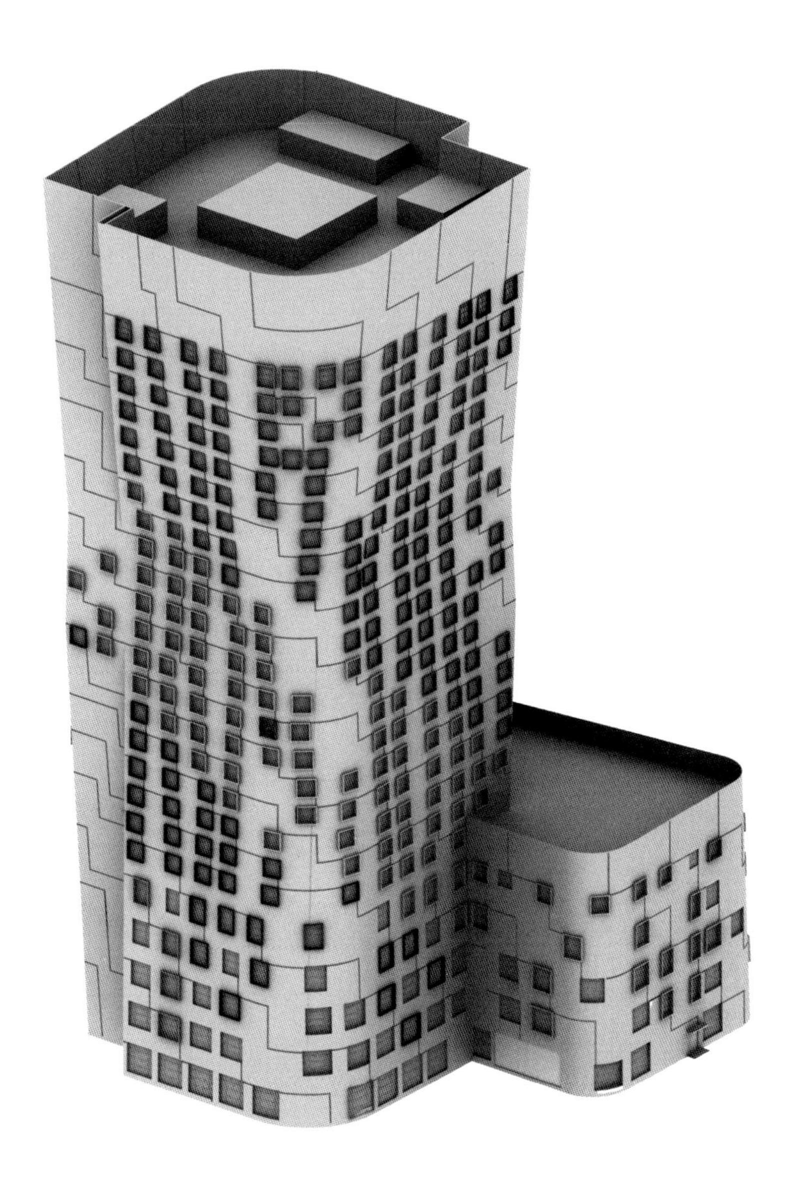

pollution and energy consumption gradually moved away from the city. Since the relocation of the Shougang Group and the termination of mining industry, the environment of Beijing City has been remarkably improved. Following the extension of Chang'an Street to the west and the openings of two high-speed freeways - Lianshi Road and Fushi Road, the connection between Mentougou District and greater Beijing area is tightening.

The constructing Oriental Bote Hotel is located on the west bound of Chang'an Street, near the Shimenying turnabout; it is the entrance from the greater Beijing area to the Mentougou District. The people's impression of Mentougou is polluted and massively underdeveloped than other locations. In order to reveal the brand new opportunities and developments of

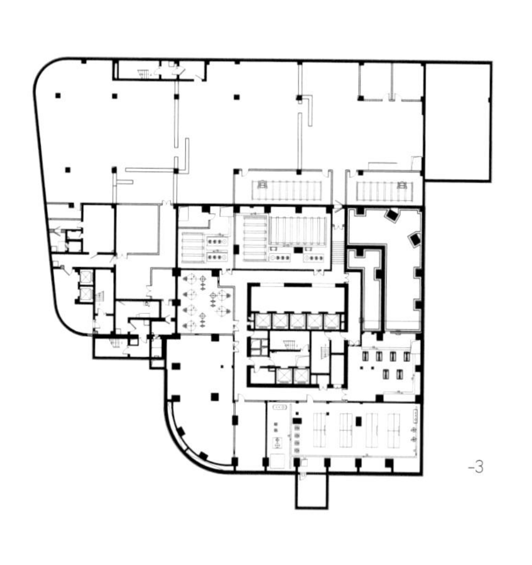
-3

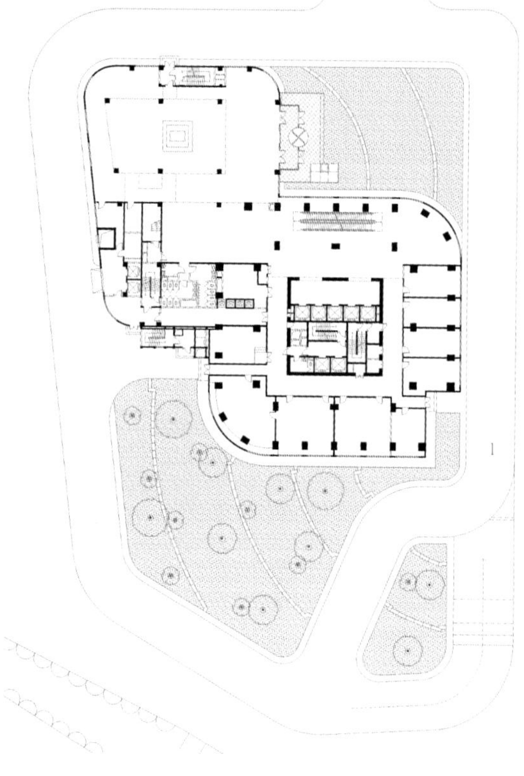
1

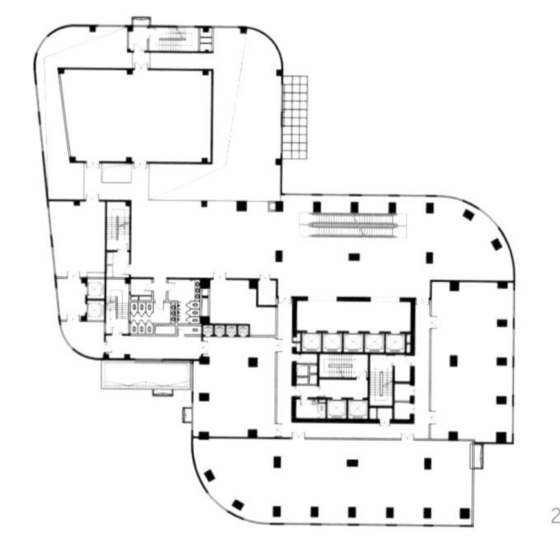
2

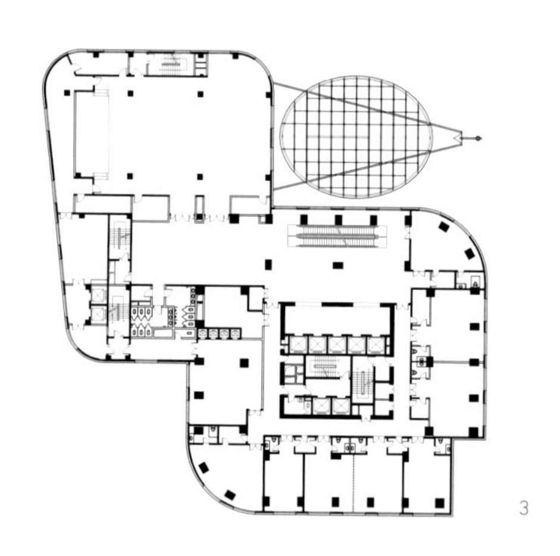
3

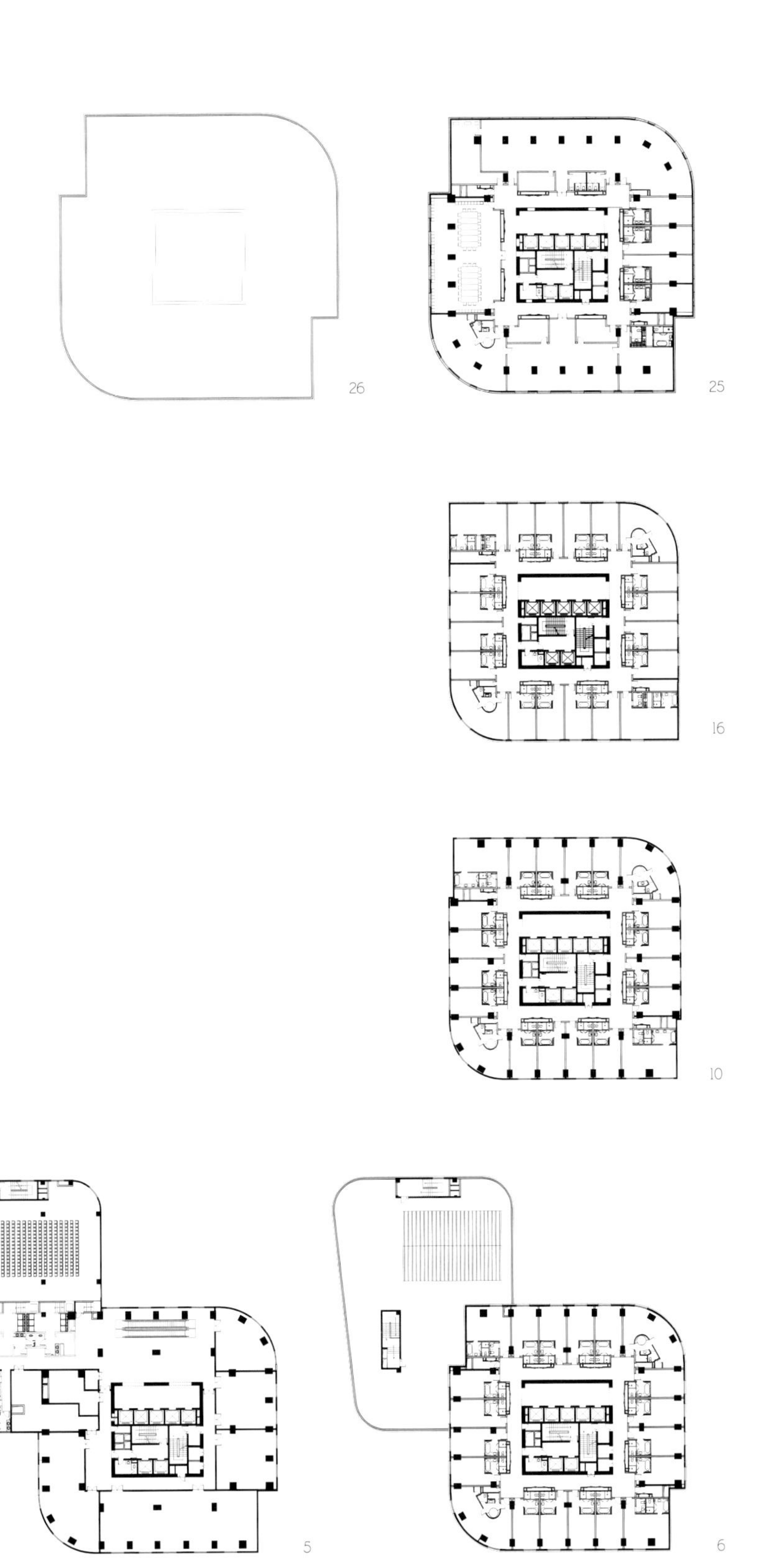
26
25
16
10
4
5
6

立面图
Facades

the Mentougou District and castaway obsolete impressions, the project adopts to fashionable modern designs. By forging new and original characteristics, the architectures gain new and original images. However, this image is not entirely forged out of the void; it matches the local environments and the atmosphere of the lost steel industry perfectly.

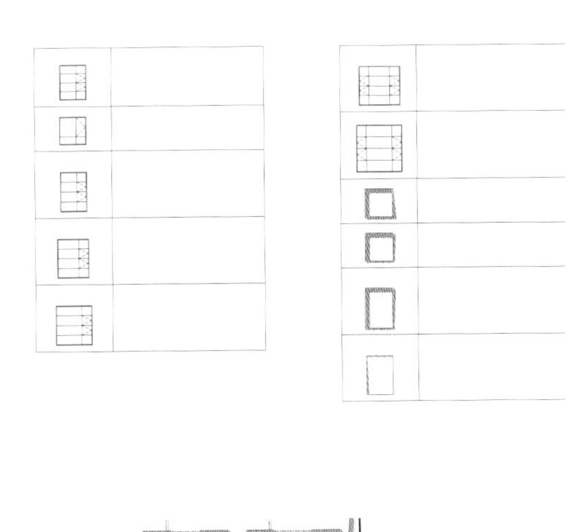

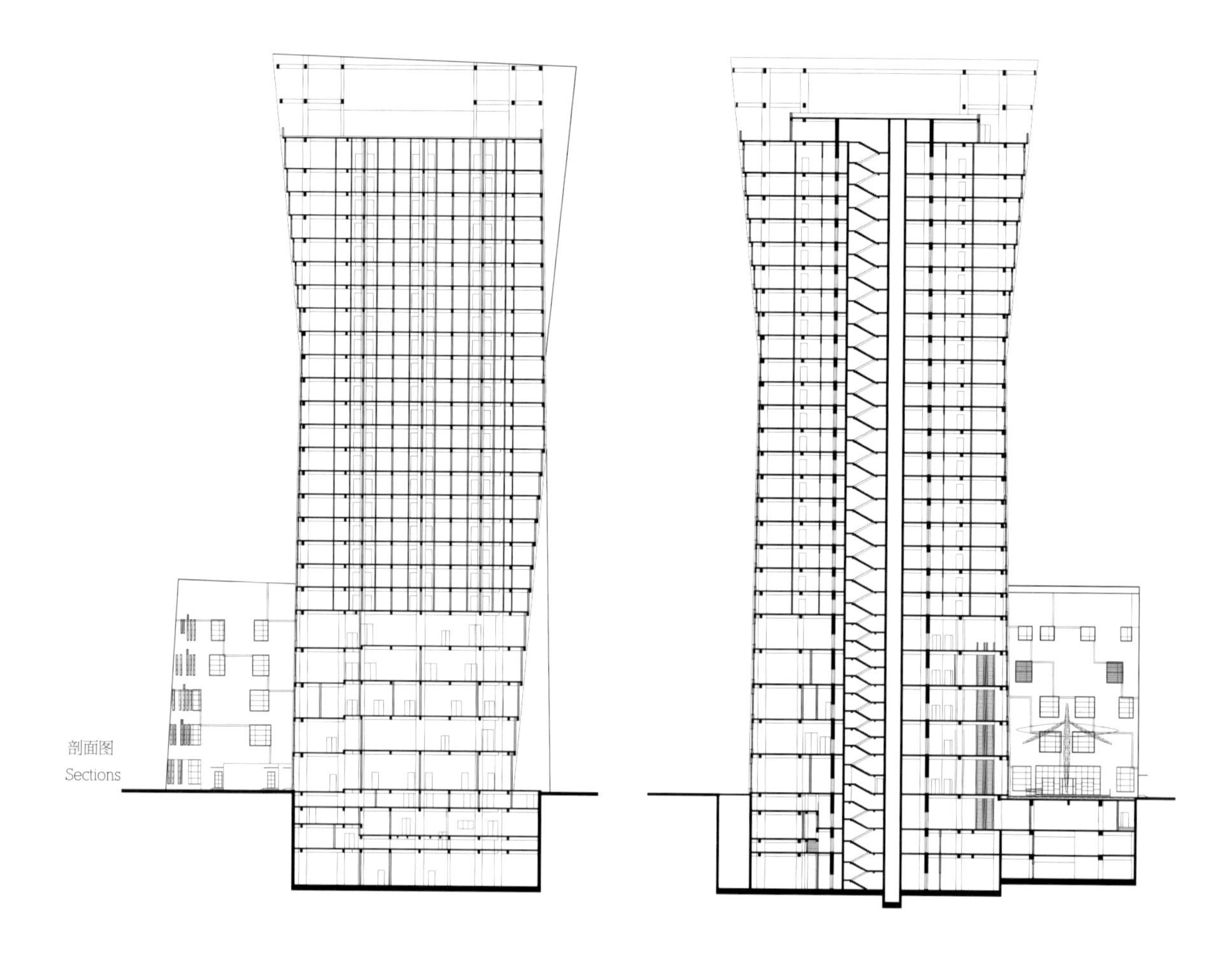

剖面图
Sections

作品索引
Work Index

黄龙瑟尔嵯国际大酒店

Huanglong Seercuo International Hotel

地点：四川 规模：22185 平方米 设计：2004 建成：2011

Location: Sichuan Size: 22185 sq. m Design: 2004 Built: 2011

设计团队：王毅、李亮、邵建、刘峰

Project Team: Wang Yi, Li Liang, Shao Jian, Liu Feng

九寨沟喜来登国际大酒店

Jiuzhaigou Sheraton International Hotel

地点：四川 规模：62800 平方米 设计：1999 建成：2002

Location: Sichuan Size: 62800 sq. m Design: 1999 Built: 2002

设计团队：王毅、邓雪娴、华夫荣、陈林、赵岚、邵建、刘峰

Project Team: Wang Yi, Deng Xuexian, Hua Furong, Chen Lin, Zhao Lan, Shao Jian, Liu Feng

呼伦贝尔民族文化园

Hulun Buir Cultural Park of Nationalities

地点：海拉尔 规模：地段 100 公顷 设计：2010

Location: Hailaer Size: 100 ha Design: 2010

设计团队：王毅、农冬红、史逸、刘啸、李磊

Project Team: Wang Yi, Nong Donghong, Shi Yi, Liu Xiao, Li Lei

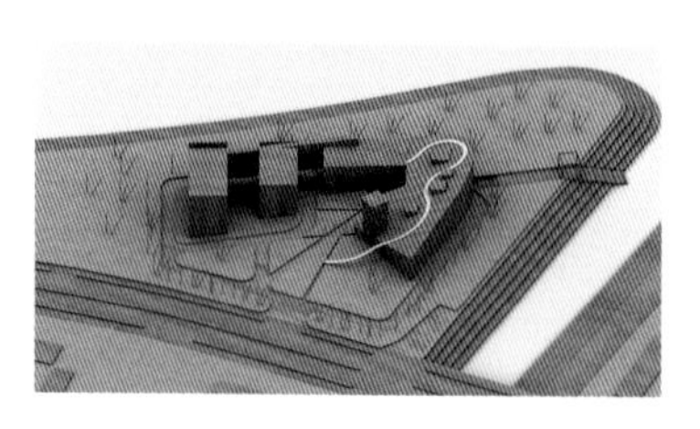

门头沟区规划展览馆

Planning Exhibition Hall of Mentougou

地点：北京 规模：19500 平方米 设计：2012

Location: Beijing Size: 19500 sq. m Design: 2012

设计团队：王毅、郭炳峰、郁美江

Project Team: Wang Yi, Guo Bingfeng, Yu Meijiang

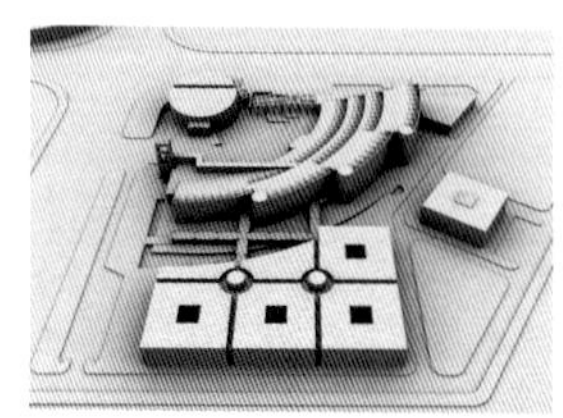

华人艺术宫

Chinese Art Palace

地点：天津 规模：25000 平方米 设计：1991

Location: Tianjin Size: 25000 sq. m Design: 1991

设计团队：王毅

Project Team: Wang Yi

御泉湾会所

Yuquanwan Club

地点：黄山 规模：1355 平方米 设计：2008 建成：2009

Location: Huangshan Size: 1355 sq. m Design: 2008 Built: 2009

设计团队：王毅、史逸

Project Team: Wang Yi, Shi Yi

北京电影学院逸夫影视艺术中心
Beijing Film Academy's Yifu Film & TV Art Center
地点：北京　规模：8168 平方米　设计：1997　建成：2000
Location: Beijing　Size: 8168 sq. m　Design: 1997　Built: 2000
设计团队：王毅、邓雪娴、邹军
Project Team: Wang Yi, Deng Xuexian, Zou Jun

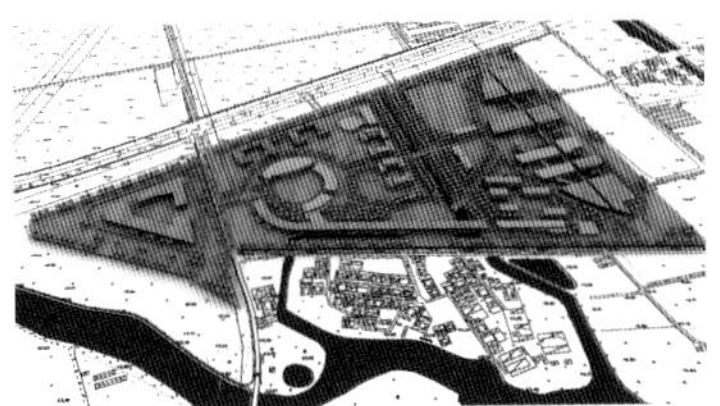

中国纪检监察学院
China College of Supervision
地点：北京　规模：36.98 公顷　设计：2008
Location: Beijing　Size: 36.98 ha　Design: 2008
设计团队：王毅、胡建新、唐鸿俊、张冰冰
Project Team: Wang Yi, Hu Jianxing, Tang Hongjun, Zhang Bingbing

大小马站历史街区
Daxiao Mazhan Historical District
地点：广州　规模：5.46 公顷　设计：2011
Location: Guangzhou　Size: 5.46 ha　Design: 2011
设计团队：王毅、农冬红、郁美江、郭炳峰、刘达
Project Team: Wang Yi, Nong Donghong, Yu Meijiang, Guo Bingfeng, Liu Da

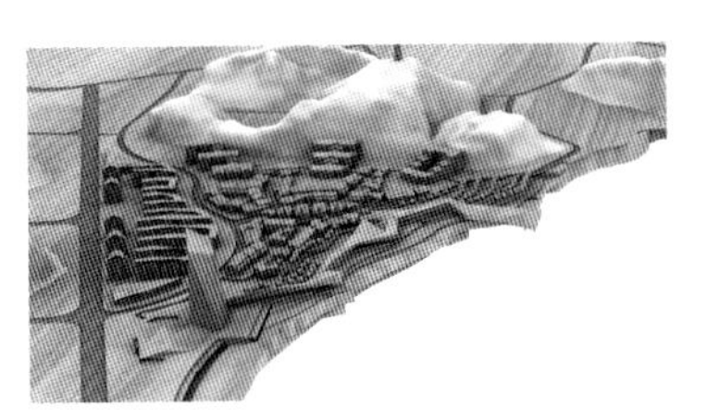

窑埠古镇
Yaobu Ancient Town
地点：柳州　规模：21.9 公顷　设计：2009
Location: Liuzhou　Size: 21.9 ha　Design: 2009
设计团队：王毅、史逸、李梓华
Project Team: Wang Yi, Shi Yi, Li Zihua

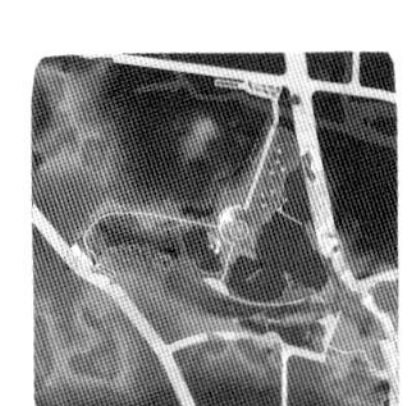

青龙山生态公园
Qinglong Mountain Eco-park
地点：宜兴　规模：53.09 公顷　设计：2011
Location: Yixing　Size: 53.09 ha　Design: 2011
设计团队：王毅、郁美江、郭炳峰
Project Team: Wang Yi, Yu Meijiang, Guo Bingfeng

文昌东楼
Wenchang East Building
地点：柳州　规模：29422 平方米　设计：2010
Location: Liuzhou　Size: 29422 sq. m　Design: 2010
设计团队：王毅、郁美江、郭炳峰、刘达
Project Team: Wang Yi, Yu Meijiang, Guo Bingfeng, Liu Da

成都南站概念方案
Conceptual Design of Chengdu South Railway Station
地点：成都　规模：14000 平方米　设计：2009
Location: Chengdu　Size: 14000 sq. m　Design: 2009
设计团队：王毅、高林、马光远、陈东旭
Project Team: Wang Yi, Gao Lin, Ma Guangyuan, Chen Dongxu

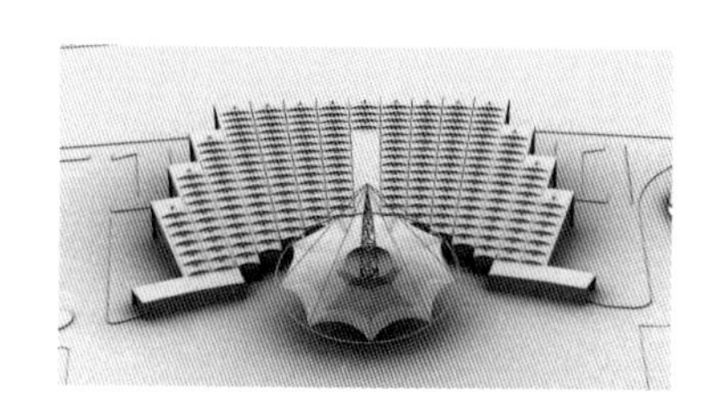

东兴国际展销中心
Dongxing International Exhibition Center
地点：东兴 规模：19000 平方米 设计：2002
Location: Dongxing Size: 19000 sq. m Design: 2002
设计团队：王毅、李亮、李华东
Project Team: Wang Yi, Li Liang, Li Huadong

文县横丹学校
Wen County's Hengdan School
地点：甘肃 规模：5451 平方米 设计：2009 建成：2010
Location: Gansu Size: 5451 sq. m Design: 2009 Built: 2010
设计团队：王毅、李梓华、史逸
Project Team: Wang Yi, Li Zihua, Shi Yi

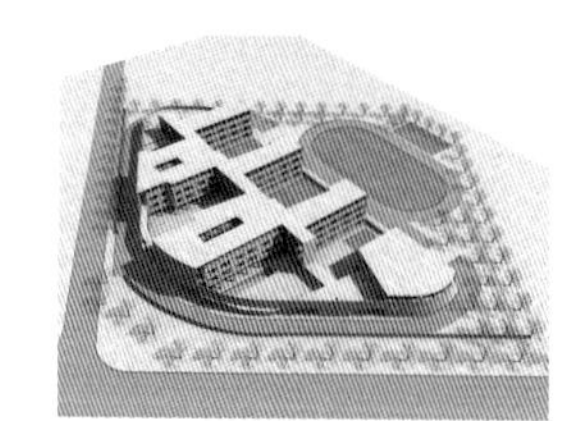

古亭山小学
Gutingshan Primary School
地点：柳州 规模：14709 平方米 设计：2010
Location: Liuzhou Size: 14709 sq. m Design: 2010
设计团队：王毅、郭炳峰、郁美江、刘达
Project Team: Wang Yi, Guo Bingfeng, Yu Meijiang, Liu Da

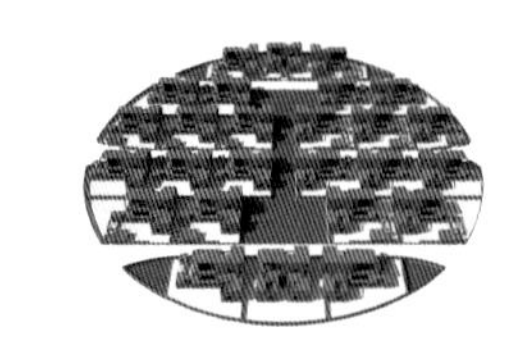

曦湾别墅
Xiwan Villa
地点：鄂尔多斯 规模：33.26 公顷 设计：2011
Location: Erdos Size: 33.26 ha Design: 2011
设计团队：王毅、郁美江、郭炳峰 合作团队：都市实践
Project Team: Wang Yi, Yu Meijiang, Guo Bingfeng Collaboration Team: Urbanus

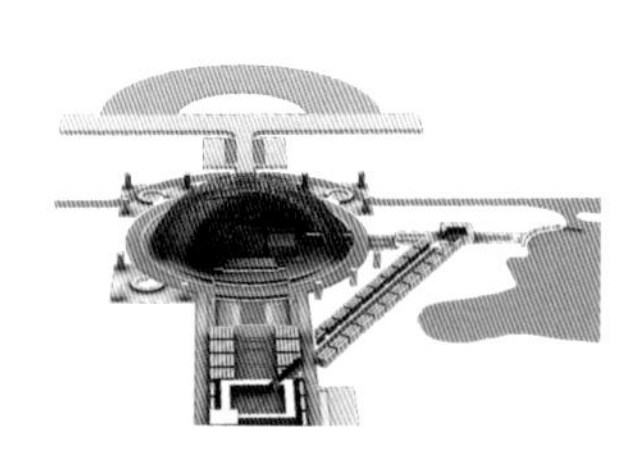

中华世纪门
China Arch for New Millennium
地点：北京 规模：52 公顷 设计：2000
Location: Beijing Size: 52 ha Design: 2000
设计团队：王毅、程晓青、雷晓明、王鹏
Project Team: Wang Yi, Cheng Xiaoqing, Lei Xiaoming, Wang Peng

曹妃甸 AB 地块概念设计
Conceptual Design of Caofeidian AB Plots
地点：曹妃甸 规模：154 公顷 设计：2010
Location: Caofeidian Size: 154 ha Design: 2010
设计团队：王毅、农冬红、史逸、韩冰浩、徐连柱
Project Team: Wang Yi, Nong Donghong, Shi Yi, Han Binghao, Xu Lianzhu

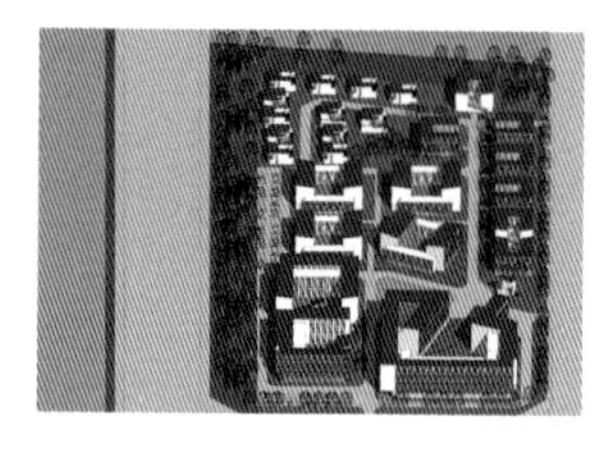

曹妃甸 B3 社区
Caofeidian B3 Community
地点：曹妃甸 规模：8 公顷 设计：2010
Location: Caofeidian Size: 8 ha Design: 2010
设计团队：王毅、史逸、马庆瑞、高宝成
Project Team: Wang Yi, Shi Yi, Ma Qingrui, Gao Baocheng

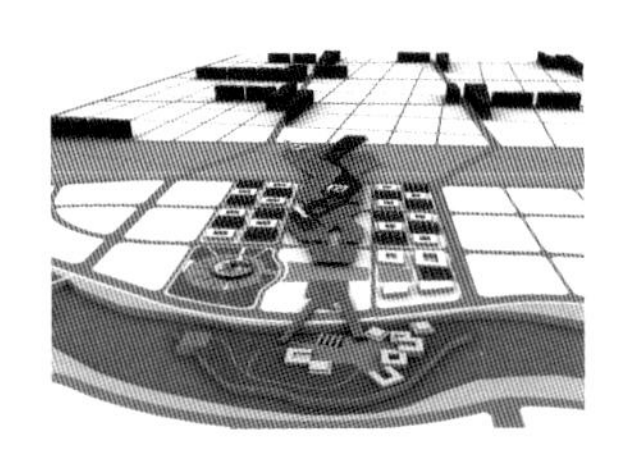

呼伦贝尔新区概念设计
Conceptual Design of Hulun Buir New District
地点：海拉尔 规模：1428.4 公顷 设计：2010
Location: Hailaer Size: 1428.4 ha Design: 2010
设计团队：王毅、杨沛儒、李梓华、史逸、刘啸、李磊
Project Team: Wang Yi, Perry Yang, Li Zihua, Liu Xiao, Li Lei

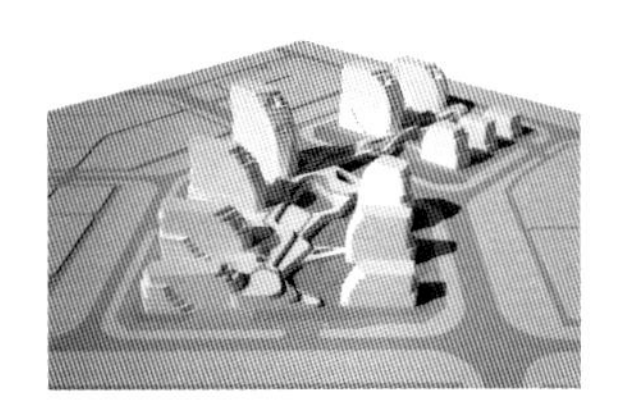

望京电子城 B12 和 C3 地块
B12 & C3 Plots of Wangjing Electronic City
地点：北京 规模：9.31 公顷 设计：2011
Location: Beijing Size: 9.31 ha Design: 2011
设计团队：王毅、高林、陈东旭、韩文飞
Project Team: Wang Yi, Gao Lin, Chen Dongxu, Han Wenfei

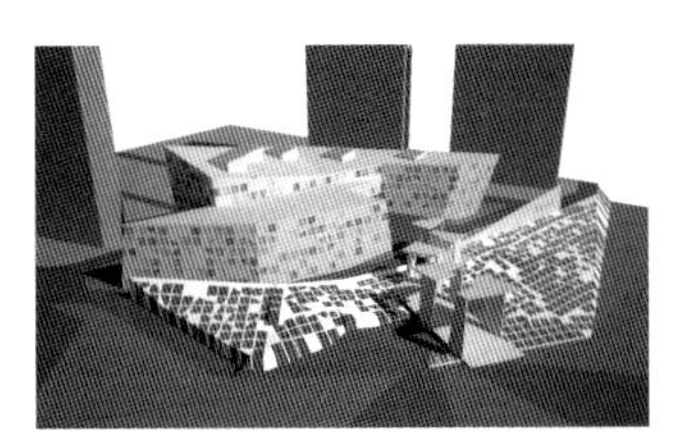

大城小院会所
Dacheng Xiaoyuan Club
地点：柳州 规模：8163 平方米 设计：2010
Location: Liuzhou Size: 8163 sq. m Design: 2010
设计团队：王毅、农冬红、郭炳峰、郁美江、刘达
Project Team: Wang Yi, Nong Donghong, Guo Bingfeng, Yu Meijiang, Liu Da

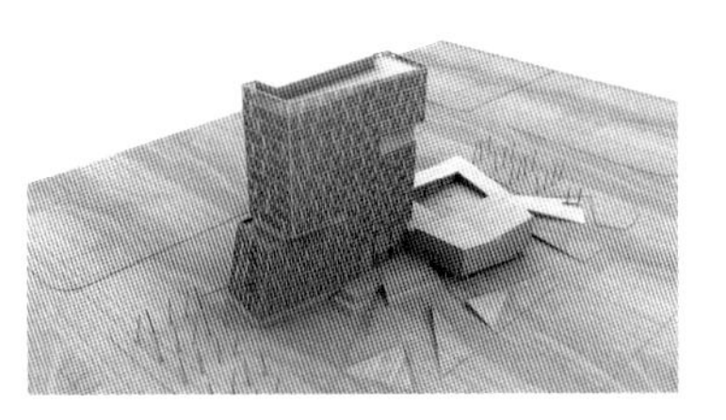

唐山市质监综合楼
Complex of Tangshan Quality Supervision
地点：唐山 规模：20510 平方米 设计：2010
Location: Tangshan Size: 20510 sq. m Design: 2010
设计团队：王毅、李梓华、史逸、
Project Team: Wang Yi, Li Zihua, Shi Yi

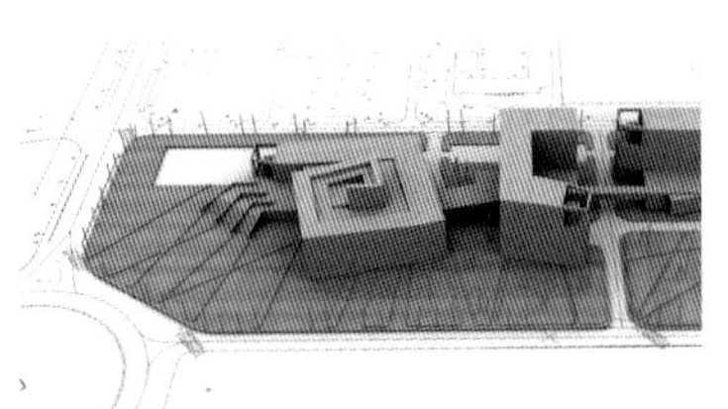

大庆规划展览馆
Planning Exhibition Hall of Daqing
地点：大庆 规模：56000 平方米 设计：2009
Location: Daqing Size: 56000 sq. m Design: 2012
设计团队：王毅、高林、韩冰浩、陈东旭

九寨沟贵宾楼水上餐厅
Water Restaurant of Jiuzhaigou Grand Hotel
地点：四川 规模：1200 平方米 设计：2012
Location: Sichuan Size: 1200 sq. m Design: 2012
设计团队：王毅、郁美江、郭炳峰
Project Team: Wang Yi, Yu Meijiang, Guo Bingfeng

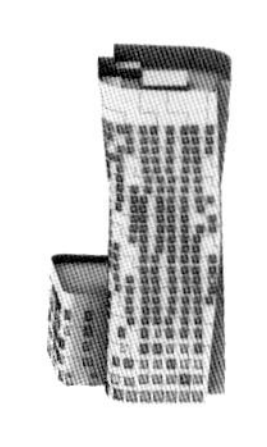

东方博特酒店（五星级）
Oriental Bote Hotel (5-star)
地点：北京 规模：47808 平方米 设计：2012 建成：在建
Location: Beijing Size: 47808 sq. m Design: 2012 Built: under construction
设计团队：王毅、高林、郭炳峰、郁美江、赵峰
Project Team: Wang Yi, Gao Lin, Guo Bingfeng, Yu Meijiang, Zhao Feng

图书在版编目(CIP)数据

本土出品 —— 王毅建筑创作札记 / 王毅著
—北京：中国建筑工业出版社，2013.9
ISBN 978-7-112-15882-9

I. ①本… II. ①王… III. ①建筑设计－作品集－中国－现代 IV. ① TU206

中国版本图书馆 CIP 数据核字 (2013) 第 222861 号

责任编辑：徐晓飞　李　鸽　施佳明

责任校对：赵　力

本土出品 —— 王毅建筑创作札记
王毅　著
*
中国建筑工业出版社出版、发行（北京西郊百万庄）
各地新华书店、建筑书店经销
北京雅昌彩色印刷有限公司制版
北京雅昌彩色印刷有限公司印刷
*
开本：787×1092 毫米　1/16　印张：$16\frac{1}{2}$　字数：330 千字
2013 年 9 月第一版　2013 年 9 月第一次印刷
定价：198.00 元

ISBN 978-7-112-15882-9

（24659）